Fundamentals
of
Soil
Ecology

Fundamentals
of
Soil
Ecology

David C. Coleman
D. A. Crossley, Jr.

Institute of Ecology
University of Georgia
Athens, Georgia

Academic Press
San Diego   New York   Boston
London   Sydney   Tokyo   Toronto

Cover art: A Cambisol profile with the organic matter well mixed in the A horizon. Because of faunal mixing there is no mineral accumulation in subsurface horizons. This is the characteristic soil of the temperate deciduous forests (from FitzPatrick, 1984; see Chapter 1 for details).

Academic Press, Inc.
A Division of Harcourt Brace & Company
525 B Street, Suite 1900, San Diego, California 92101-4495

United Kingdom Edition published by
Academic Press Limited
24-28 Oval Road, London NW1 7DX

Library of Congress Cataloging-in-Publication Data

Coleman, David C., date.
   Fundamentals of soil ecology / by David C. Coleman, D. A. Crossley, Jr.
      p.   cm.
   Includes bibliographical references and index.
   ISBN 0-12-179725-2
   1. Soil ecology. 2. Soil biology.   I. Crossley, D. A.
   II. Title
   QH541.5.S6C65   1995
   574.5'26404--dc20   95-31375
                                          CIP

PRINTED IN THE UNITED STATES OF AMERICA
95   96   97   98   99   00   MM   9   8   7   6   5   4   3   2   1

Contents

3 | *Secondary Production: Activities of Heterotrophic Organisms—Microbes*

7 | *Future Developments in Soil Ecology*

Preface

One of the last great frontiers in biological and ecological research is the soil. Civilizations, so dependent on soils as a source of nutrients for food, owe them a considerable debt.

Over the span of several millennia, there has been concern about the use and misuse of soils. There is ample evidence that numerous civilizations, from ancient Sumeria and Babylonia to those that use modern high-intensity, high-input production agriculture, have suffered with problems of long-term sustainability (Whitney, 1925; Pesek, 1989).

Indeed, one prominent soil physicist has been moved to comment that "the plow has caused more destruction to civilizations than the sword" (Hillel, 1991). Perhaps the adage of "beating swords into plowshares" needs rethinking. As we will discover in the course of this book, it is truly a time to be working "with nature" and to cease treating soils as a "black box."

Soil is a unique entity. It has its origins in physical, chemical, and biological interactions between the parent materials and the atmosphere. The simplest definitions of soil follow the common understanding, such as "the upper layer of earth which can be dug or plowed and in which plants grow" (*Webster's New Collegiate Dictionary*). The soil scientist defines it as "a natural body, synthesized in profile form from a variable mixture of broken and weathered minerals and decaying organic matter, which covers the earth in a thin layer and which supplies, when containing the proper amounts of air and water, mechanical support and, in part, sustenance for plants" (Buckman and Brady, 1970). This definition recognizes that soil has vertical structure, is composed of a variety of materials, and has a biological nature as well—it is derived in part from decaying organic matter. Nevertheless, uncertainties emerge when more restrictive definitions are attempted. How deep is soil or when is nonsoil encountered? Working definitions of soil depth range from 1 m to many, depending on the ecosystem and the nature of the investigations. Are barren, rocky areas excluded if they do not allow growth of higher plants? Lindeman (1942) considered the substratum of a lake as a benthic soil (see also Jenny, 1980). When do simple sediments become soil? When they can support plant growth? Only after biological, physical, and chemical interactions convert sediments into an organized profile?

Soils are composed of a variable combination of four key constituents: minerals, organic matter, water, and air. Of the Greco-Roman concepts of fundamental constituents, earth, air, fire, and water, three of the four are contained within the broad concept of soil. Indeed, if the energetic processes of life ("the fire of life" (Kleiber, 1961)) are included within soil, then all four of the ancient "elements" are present therein.

Are living organisms part of soil? We would include the phrase "with its living organisms" in the general definition of soil. Thus, from our viewpoint soil is alive and is composed of living and nonliving components, having many interactions.

It is as a part of that larger unit, the terrestrial ecosystem, that soil must be studied and conserved. The interdependence of terrestrial vegetation and animals, soils, atmosphere, and hydrosphere is complex, with many feedback mechanisms. When we view the soil system as an environment for organisms, we must remember that the biota have been involved in its creation, as well as adapting to life within it. The principles by which organisms in soils are distributed, interact, and carry on their lives are far from completely known, and the importance of the biota for soil processes is not often appreciated.

This book on soil ecology emphasizes the interdisciplinary nature of studies in ecology as well as soils. Considerable "niche overlap" (similarity in what they do, i.e., their "profession" (Elton, 1927)) exists between the two disciplines of ecology and soil science. Ecology, which is heavily organism-oriented, is concerned with all forms of life in relation to their environment. Soil science, in contrast, contains several other aspects in addition to soil biology, such as soil genesis and classification, soil physics, and soil chemistry. A broader view of ecology asks, "How do systems work?" From that perspective, ecology and soil science share similar objectives.

The overlap between ecology and soil science is both extensive and interesting. Aspects of soil physics, chemistry, and mineralogy have a great impact on how many different kinds of organisms coexist in the opaque, complex, semiaquatic milieu that we call soil. We first describe what soils are and how they are formed, and then discuss some of the current research being done in soil ecology.

With a rising tide of interest worldwide in soils, and in belowground processes in general, numerous types of studies using tools in all ranges of the size and electromagnetic energy spectra and encompassing from microsites to the biosphere are now possible. Significant achievements during the past 5 to 10 years make a book of this sort both timely and useful. This book is intended primarily as a source of ideas and concepts and thus is intended as a supplemental reference for courses in ecology, soil science, and soil microbiology.

We hope that we will interest a new generation of ecologists and soil scientists in the world of soil ecology: the interface between biology, chemistry, and physics of soil systems.

Acknowledgments

We dedicate this book to several people who contributed materially to its content and tenor. First, our graduate students and postdoctorals at both Colorado State University and the University of Georgia served to sharpen our focus in the classroom and in less-formal "symposia." We also thank our many colleagues, including Paul Hendrix, John Blair, Ted Elliott, Liam Heneghan, Rob Parmelee, and Dave Wright, for stimulating advice and counsel. We have modeled our approach on that of our mentor, Eugene P. Odum, whose textbooks have been an inspiration to us both. Last but not least, we thank Linda Slaney for her many hours of support on the manuscript (particularly the visuals) and our wives, Fran and Dot, for their forbearance in the final throes of preparing this book.

1 | *Introduction: The Fitness of the Soil Environment*

WATER AS A CONSTITUENT OF SOIL

> The occurrence of water is, moreover, not less important and hardly less general upon the land. In addition to lakes and streams, water is almost everywhere present in large quantities in the soil, retained there mainly by capillary action, and often at greater depths. (*Henderson, 1913*)

Lawrence J. Henderson, a noted physical chemist and physiologist, published "The Fitness of the Environment" (1913) which was a landmark among books on biological topics. Henderson's thesis is that one substance, water, is responsible for the characteristics of life and the biosphere as we know it. The highly bipolar nature of water, with its twin hydrogen bonds, leads to a number of intriguing characteristics, e.g., high specific heat, which have enabled life in the thin diaphanous veil of the biosphere (Lovelock, 1979, 1988) to extend and proliferate, almost endlessly.

A central fact of soil science is that certain physicochemical relationships of matter in all areas of the biosphere are mediated by water. Thus soil, which we normally think of as opaque and solid, from the wettest organic muck soil, to the parched environs of the Kalahari, Gobi, or Mojave deserts, is dominated by the amount and availability of water.

Consider water in each of its phases: solid, liquid, and gaseous.

1. *Solid.* In aquatic ecosystems, water freezes from the top down, as it has its greatest density at 4°C. This allows for organismal activity to continue at lower depths, and in sediments as well. In soil, the well-insulated nature of the soil materials and water with its high specific heat means that there is less likelihood of rapid freezing. Water expands when it freezes. In more polar climates (and in some temperate ones), soil can be subjected to "frost heaving," which can be quite disruptive, depending on the nature of the subsurface materials.

2. *Liquid.* Water's high specific heat of one calorie per gram per degree Celsius increase in temperature has a significant stabilizing influence in bodies of water and soil (Table 1.1). The effect of high specific heat is to reduce fluctuations in temperature. The location of the liquid, in various films or in empty spaces, has a marked influence on the soil biota.

3. *Vapor.* It is somewhat counterintuitive but true that the atmosphere within air-dry soil (gravimetric water content of 2% by weight) has a relative humidity of 98%. The consequences of this humidity for life in the soil are very profound. Most soil organisms spend their lives in an atmosphere saturated with water. Many soil animals absorb and lose water through their integuments, and are entirely dependent on saturated atmospheres for their existence.

From the pragmatic viewpoint of the soil physicist, we can consider aqueous and vapor phases of water conjointly. Following a moisture–release curve, one can trace the pattern of water, in volume and

TABLE 1.1.   Specific Heats [a] of Various Substances [b]

Substance	Specific heat
Lead	0.03
Iron	0.10
Quartz	0.19
Sugar	0.30
Chloroform	0.24
Hexane	0.50
Water	
Liquid	1.0
Solid	0.5
Gas	0.3–0.5
Ammonia, liquid	1.23

[a]Calories (= 4.18 J) to raise 1 g by 1°C.
[b]Modified from Hadas (1979).

location in the soil pore spaces, in the following manner (Vannier, 1987; 1981). Starting with free-standing or gravitational water at saturation, the system is essentially subaquatic (Fig. 1.1). With subsequent evaporation from the soil, the free-standing water disappears, leaving some capillary bound water (Fig. 1.2), which has been termed the edaphic system. Further evaporation then occurs, resulting in a virtual absence of any capillary water, leaving only the adsorbed water, at a very high negative water tension (Fig. 1.3).

The implications of this complex three-dimensional milieu are of fundamental importance for a very diverse biota. Vannier (1973) proposed the term "porosphere" for this intricate arrangement of sand, silt, clay, and organic matter. Primitive invertebrates first successfully undertook the exploitation of aerial conditions at the beginning of the Paleozoic era (Vannier, 1987). This transition probably took place via the soil medium, which provided the necessary gradient between the fully aquatic and aerial milieus. This water-saturated environment, so necessary for such primitive wingless (Apterygote) forms as the Collembola or springtails (Fig. 1.1), is equally important for the transient life-forms such as the larval forms of many flying insects, such as Diptera and Coleoptera. In addition, many of the micro- and mesofauna, described in Chapter 4, could be considered part of the "terrestrial nannoplankton" (Stout, 1963). Stout included all of the water-film inhabitants, namely: bacteria and yeasts,

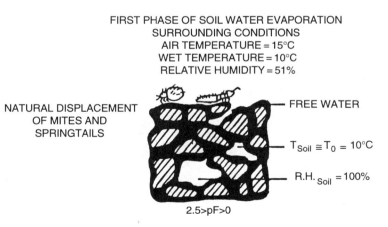

FIRST PHASE OF SOIL WATER EVAPORATION
SURROUNDING CONDITIONS
AIR TEMPERATURE = 15°C
WET TEMPERATURE = 10°C
RELATIVE HUMIDITY = 51%

NATURAL DISPLACEMENT
OF MITES AND
SPRINGTAILS

FREE WATER

$T_{Soil} \cong T_0 = 10°C$

$R.H._{Soil} = 100\%$

2.5>pF>0

FIGURE 1.1 Gravitational moisture (the subaquatic system) in the soil framework; pF, –log cm H_2O suction; R.H., relative humidity. Reprinted, with permission, from Vannier, G. (1987). The porosphere as an ecological medium emphasized in Professor Ghilarov's work on soil animal adaptations. *Biol. Fertil. Soils* **3**, 39–44. Copyright 1984 Springer-Verlag.

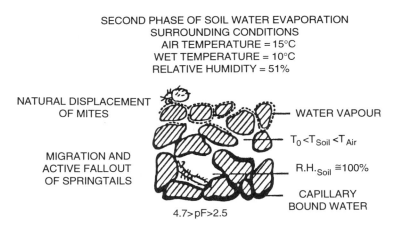

SECOND PHASE OF SOIL WATER EVAPORATION
SURROUNDING CONDITIONS
AIR TEMPERATURE = 15°C
WET TEMPERATURE = 10°C
RELATIVE HUMIDITY = 51%

NATURAL DISPLACEMENT
OF MITES

WATER VAPOUR

$T_0 < T_{Soil} < T_{Air}$

MIGRATION AND
ACTIVE FALLOUT
OF SPRINGTAILS

R.H.$_{Soil}$ ≅100%

CAPILLARY
BOUND WATER

4.7 > pF > 2.5

FIGURE 1.2 Capillary moisture (the edaphic system) in the soil framework. Reprinted, with permission, from Vannier, G. (1987). The porosphere as an ecological medium emphasized in Professor Ghilarov's work on soil animal adaptations. *Biol. Fertil. Soils* **3**, 39–44. Copyright 1984 Springer-Verlag.

protozoa, rotifers, nematodes, copepods, and microdrili such as enchytraeids, also called potworms. Raoul Francé, a German sociologist, made analogies between aquatic plankton and the small and medium-sized organisms which inhabit the water films and water-filled pores in soils, terming them "Das Edaphon" (Francé, 1921).

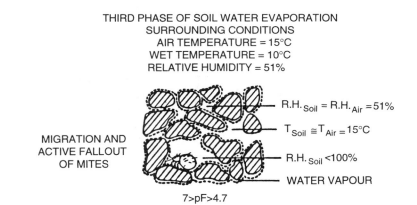

THIRD PHASE OF SOIL WATER EVAPORATION
SURROUNDING CONDITIONS
AIR TEMPERATURE = 15°C
WET TEMPERATURE = 10°C
RELATIVE HUMIDITY = 51%

R.H.$_{Soil}$ = R.H.$_{Air}$ = 51%

$T_{Soil} \cong T_{Air} = 15$°C

MIGRATION AND
ACTIVE FALLOUT
OF MITES

R.H.$_{Soil}$ < 100%

WATER VAPOUR

7 > pF > 4.7

FIGURE 1.3 Adsorptional moisture (the aerial system) in the soil framework. Reprinted, with permission, from Vannier, G. (1987). The porosphere as an ecological medium emphasized in Professor Ghilarov's work on soil animal adaptations. *Biol. Fertil. Soils* **3**, 39–44. Copyright 1984 Springer-Verlag.

As noted in Figs. 1.1–1.3, there is a marked difference in moisture requirements of some of the soil microarthropods. Thus another major group, the Acari or mites, are often able to tolerate considerably more desiccation than the more sensitive Collembola. In both cases, the microarthropods are gradually expelled from the soil matrix as the desiccation sequence just described continues.

Other organisms, more dependent on the existence of free water or water films, include the protozoa and nematoda; the life histories and feeding characteristics of which are covered in Chapter 4. In a sense, the very small fauna, and the bacteria they feed upon, exist in a qualitatively different world from the other fauna, or fungi, which move into and out of various water films and through various pores which are less than 100% saturated with water vapor, with comparative ease (Hattori, 1994).

In conclusion, this overview of soil physical characteristics and their biological consequences notes the following: "For a physicist, porous bodies are solids with an internal surface which endows them with a remarkable set of hygroscopic properties. For example, a clay such as bentonite has an internal surface in excess of 800 m^2g^{-1}, and a clay soil containing 72% montmorillonite possesses an internal surface equal to 579 m^2g^{-1}. The capacity to condense gases on free walls of capillary spaces (the phenomenon of adsorption) permits porous bodies to reconstitute water reserves from atmospheric water vapor" (Vannier, 1987). Later, we will address the phenomenon of adsorption in other contexts, ones which are equally important for soil functioning as we know it.

ELEMENTAL CONSTITUTION OF SOIL

Many elements are found within the earth's crust, and most of them are in soil as well. However, a few elements predominate. These are hydrogen, carbon, oxygen, nitrogen, phosphorus, sulfur, aluminum, silicon, and alkali and alkaline earth metals. Various trace elements, also called micronutrients, are present as enzyme cofactors and include iron, cobalt, nickel, copper, magnesium, manganese, molybdenum, and zinc.

A more functional and esthetically pleasing approach is to define soil as predominantly a sand–silt–clay matrix, containing living (biomass) and dead (necromass) organic matter, with varying amounts of gases and liquids within the matrix. In fact, the interactions of geological, hydrological, and atmospheric (Fig. 1.4) facets overlap with those of the biosphere, leading to the union of all, overlapping, in part, in the pedosphere. Soils, in addition to the three geometric

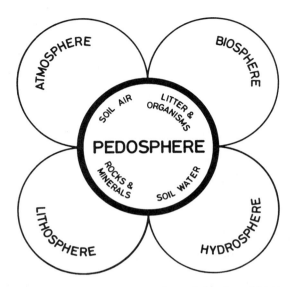

FIGURE 1.4   The pedosphere, showing interactions of abiotic and biotic entities in the soil matrix (from FitzPatrick, 1984).

dimensions, are also greatly influenced by the fourth dimension: time, over which the physicochemical and biological processes occur.

HOW SOILS ARE FORMED

Soils are the resultant of the interactions of several factors: climate, organisms, parent material, and topology (relief), all acting through time (Jenny, 1941, 1980) (Fig. 1.5). These factors affect major ecosystem processes, e.g., primary production, decomposition, and nutrient cycling, which lead to the development of ecosystem properties unique to that soil type, given its previous history. Thus such characteristics as cation-exchange capacity, texture, structure, and organic matter status are the outcome of the aforementioned processes operating as constrained by the controlling factors. Different arrays of processes may predominate in various ecosystems (Fig. 1.5).

PROFILE DEVELOPMENT

The abiotic and biotic factors noted earlier lead to certain chemical changes down through the top few decimeters of soil (Figs. 1.6a and 1.6b). In many soils, particularly in more mesic or moist regions of the

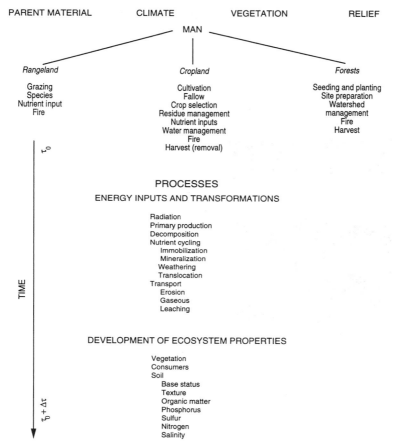

FIGURE 1.5 Soil-forming factors and processes, and the interaction over time (1980) (from Coleman *et al.*, 1983, modified from Jenny, 1980).

world, there is leaching and redeposition of minerals and nutrients, often accompanied by a distinct color change (profile development). Thus, as one descends through the profile from the air-litter surface, one passes through the litter, fermentation, and humification zones (a_0, a_i, and a_{ii}), respectively, then reaching the mineral soil surface, which contains the preponderant amount of organic matter (A horizon). The B horizon is next, with deeper-dwelling organisms and somewhat weathered material. This is followed by the C horizon, the unconsolidated mineral material above bedrock. More details on soil

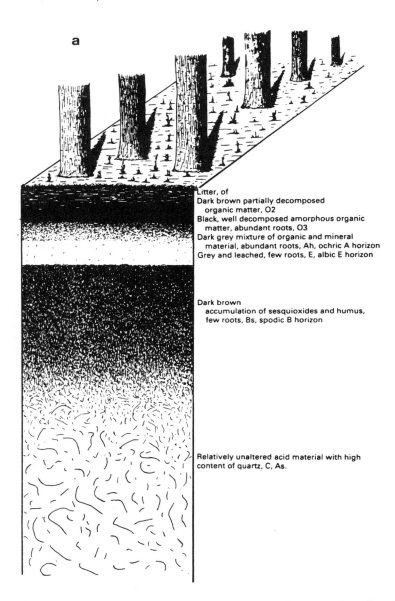

a

Litter, of
Dark brown partially decomposed
    organic matter, O2
Black, well decomposed amorphous organic
    matter, abundant roots, O3
Dark grey mixture of organic and mineral
    material, abundant roots, Ah, ochric A horizon
Grey and leached, few roots, E, albic E horizon

Dark brown
    accumulation of sesquioxides and humus,
    few roots, Bs, spodic B horizon

Relatively unaltered acid material with high
content of quartz, C, As.

FIGURE 1.6   (a) A Podzol (spodosol in North American soil taxonomy) profile with minerals accumulating in subsurface horizons. This is the characteristic soil of coniferous forests. (b) A Cambisol profile with the organic matter well mixed in the A horizon. Because of faunal mixing there is no mineral accumulation in subsurface horizons. This is the characteristic soil of the temperate deciduous forests (from FitzPatrick, 1984).

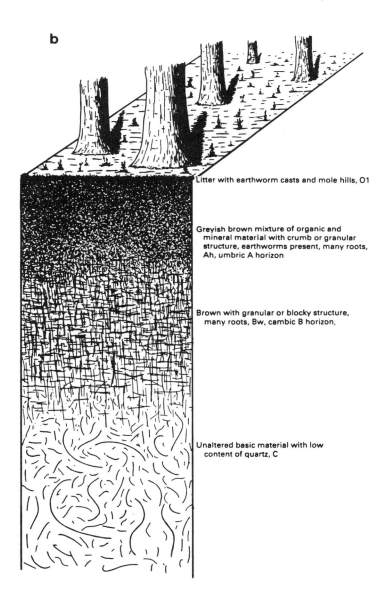

b

Litter with earthworm casts and mole hills, O1

Greyish brown mixture of organic and
mineral material with crumb or granular
structure, earthworms present, many roots,
Ah, umbric A horizon

Brown with granular or blocky structure,
many roots, Bw, cambic B horizon,

Unaltered basic material with low
content of quartz, C

classification and profile formation are given in soils textbooks, such
as Russell (1973) and Brady (1974).

The work of the soil ecologist is made somewhat easier by the fact
that the top 10–15 cm of the A horizon and the L, F, + H horizons (or

a_0, a_i, and a_{ii} in North American terminology) of forested soils have the majority of plant roots, microbes, and fauna (Coleman *et al.*, 1983; Paul and Clark, 1989), and hence a majority of the biological and chemical activities occur in this layer. Indeed, a majority of microbial and algal-feeding fauna, such as protozoa (Elliott and Coleman, 1977; Kuikman *et al.*, 1990), rotifers, and tardigrades (Leetham *et al.*, 1982), are in the surface 1 or 2 cm. Microarthropods are most abundant in the top 5 cm of forest soils (Schenker, 1984) or grassland soils (Seastedt, 1984a). This region may be "primed," in a sense, by the continual input of leaf, twig, and root materials, as well as algal and cyanobacterial production and turnover in some ecosystems whereas soil mesofauna such as nematodes and microarthropods may be concentrated in the top 5 cm. Significant numbers of nematodes may be found at several meters' depth in xeric sites, such as deserts in the American southwest (Freckman and Virginia, 1989).

SOIL TEXTURE

Historically, texture was a term used to describe the workability of an agricultural soil. A heavy clay soil required more effort (horsepower) to till than a lighter, sandy loam (Russell, 1973). A more quantifiable approach is to characterize soils in terms of the sand, silt, and clay present, which are ranged on a spectrum of light–intermediate–heavy, or sandy–silt–clay. The array of textural classes (Fig. 1.7) shows percentages of sand, silt, and clay, and the resulting soil types, such as sandy, loamy, or clayey soils.

The origin and mineralogical composition of mineral particles in soil is a most interesting and complex one. The particles are in two major categories: (1) crystalline minerals derived from primary rock, and (2) those derived from weathering animal and plant residues. The microcrystalline forms are composed of calcium carbonate, iron or aluminum oxides, or silica.

The clay fraction, so important in imparting specific physical properties to soils, to microbial life, and to plant activity via nutrient availability, is composed of particles less than 2 μm in diameter. Unlike the sand/silt minerals, clays are weathered forms of primary minerals, and hence they are referred to as secondary minerals. Coarse clay particles (0.5 μm) often are derived from quartz and mica; finer clays (0.1 μm) are clay minerals or weathered products of these (such as hydrated ferric, aluminum, titanium, and manganese oxides).

The roles of coarse and fine clays in organic matter dynamics are under intensive scrutiny in several laboratories around the world (Anderson and Coleman, 1985; Oades and Waters, 1991). It is possible

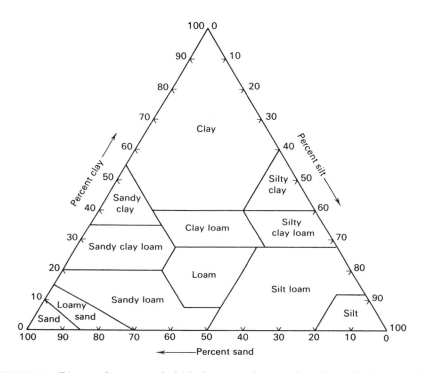

FIGURE 1.7   Diagram by means of which the textural name of a soil may be determined from a mechanical analysis. In using the diagram, the points corresponding to the percentages of silt and clay present in the soil under consideration are located on the silt and clay lines, respectively. Lines are then projected inward, parallel in the first case to the clay side of the triangle and in the second case parallel to the sand side. The name of the compartment in which the two lines intersect is the class name of the soil in question (modified from Brady, 1980).

that labile, i.e., easily metabolized, constituents of organic matter are preferentially adsorbed onto fine clay particles and may be a significant source of energy for the soil microbes (Anderson and Coleman, 1985).

CLAY MINERAL STRUCTURE

The clay minerals in soil are in the class of layer-lattice minerals and are made up of sheets of hydroxyl ions or oxygen. The clay minerals fall into two groups: those with three groups of ions lying in a plane, i.e., the 1:1 group of minerals, and those with four groups of ions lying in a plane, i.e., the 2:1 group of minerals. The type mineral of the 1:1 group is kaolinite and typically has a very low charge on it. In contrast, the 2:1 type mineral, for example, illite, carries an appreciably higher negative charge per unit weight than the kaolin group. More detailed information on the clay particles, their composition, and charges on them is given in Theng (1979) and Oades *et al.* (1989).

Perhaps the main concern to the soil ecologist is the extremely high surface area found per gram of clay mineral. Surface areas can range from 50–100 m² per gram for kaolinitic clays, to 300–500 for vermiculites, to 700–800 for well-dispersed smectites (Russell, 1973). These impressively large surface areas can play a pivotal role in adsorbing and desorbing inorganic and organic constituents in soils, and have only recently been treated in an appropriately analytical fashion as an integral part of the soil nutrient system (Tisdall and Oades, 1982).

SOIL STRUCTURE

Structure refers to the ways in which soil particles are arranged or grouped spatially. The groupings may occur at any size level on a continuum from either extreme of what are nonstructural states: single grained (such as loose sand grains) or massive aggregates of aggregates (large, irregular solid).

The implications of soil structure refer not only to the particles but also extend to the pore spaces within the structure, as noted earlier. Indeed, it is the nature of the porosity which exists in a well-structured soil which leads to the most viable communities within it. This in turn has strong implications for ecosystem management, particularly for agroecosystems (Elliott and Coleman, 1988).

Several types of structural forms are found in soils. The four major types are plate-like, prism-like, block-like, and spheroidal (Fig. 1.8). All of these are "variations on a theme," as it were, of a fundamental unit of soil aggregation: the ped. A ped is a unit of soil structure, such as an aggregate, crumb, prism, block, or granule, formed by natural processes. This is distinguished from a clod, which is artificial or manmade (Brady, 1974). Soils may have peds of differing shapes, in surface and subsurface horizons. These are the result of differing temperature, moisture, chemical, and biological conditions at various levels in the soil profile.

Another concept is helpful in soil structure: the pedon. This is an area, from 1 to 10 m², under which a soil may be fully characterized. Later in the book, we will consider the arrangement of soil units in a landscape and in an entire region. Next, we will examine some of the causes for the formation, or genesis, of soil structure.

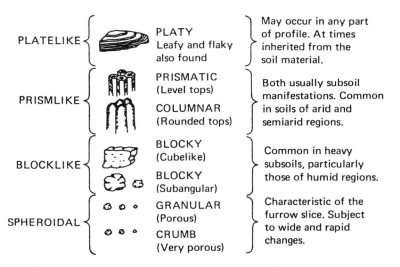

PLATELIKE — PLATY — Leafy and flaky also found — May occur in any part of profile. At times inherited from the soil material.

PRISMLIKE — PRISMATIC (Level tops) — COLUMNAR (Rounded tops) — Both usually subsoil manifestations. Common in soils of arid and semiarid regions.

BLOCKLIKE — BLOCKY (Cubelike) — BLOCKY (Subangular) — Common in heavy subsoils, particularly those of humid regions.

SPHEROIDAL — GRANULAR (Porous) — CRUMB (Very porous) — Characteristic of the furrow slice. Subject to wide and rapid changes.

FIGURE 1.8 Various structural types found in mineral soils. Their location in the profile is suggested. In arable topsoils, a stable granular structure is prized (from Brady, 1974).

Input of organic matter to soil is one of the major agents of soil structure. The organic matter comes from both living and dead sources (roots, leaves, microbes, and fauna). Various physical processes, such as deformation and compression by roots and soil fauna, and freezing/thawing, or wetting/drying, also have significant influences on soil structure. It is generally recognized that plant roots and humus (resistant organic breakdown products) play a major role in the formation of aggregates (Elliott and Coleman, 1988; Paul and Clark, 1989). However, bacteria and fungi and their metabolic products play an equally prominent role in promoting granulation (Griffiths, 1965; Cheshire, 1979; Foster, 1985). We will explore organic matter dynamics in the sections on soil biology.

The interaction of organic matter and mineral components of soils has a profound effect and influence on cation adsorption capabilities. The interchange of cations in solution with cations on these surface-active materials is an important phenomenon for soil fertility. The capacity of soils to adsorb ions (the cation-exchange capacity) is due to the sum of exchange sites on both organic matter and minerals. However, in most soils, organic matter has the higher exchange capacity (number of exchange sites). For a more extensive account see Paul and Clark (1989).

An additional aspect of aggregates, their stabilization once they are formed, is significant for soil ecology. Stabilization is the result of various binding agents. Plant and microbial polysaccharides and gums serve as binding agents (Harris et al., 1964; Cheshire, 1979; Cheshire et al., 1984). A variety of other organic compounds act as binding agents (Cheshire, 1979), and some biological agents, such as roots and fungal hyphae (Tisdall and Oades, 1979, 1982; Tisdall, 1991), play a similar role.

There is a hierarchical nature to the ways in which soil structure is achieved, and it reflects the biological interactions within the soil matrix (Elliott and Coleman, 1988). Several Australian researchers (Tisdall and Oades, 1982; Oades, 1984; Waters and Oades, 1991) have noted how the processes of structuring soils extend over many orders of magnitude, from the level of the individual clay platelet, to the ped in a given soil. For most of the biologically significant interactions, one can consider changes across a range of at least five orders of magnitude (Tisdall and Oades, 1982; Fig. 1.9). Not all soils are aggregated by biological agents; for heavily weathered Oxisols, with kaolinite-oxide clays, there seems to be no hierarchy of organization below 20 μm because only physicochemical forces predominate there (Oades and Waters, 1991). Studies in our Horseshoe Bend agroecosystem project at the University of Georgia have uncovered significant differences between tillage regimes (conventional, moldboard plowing versus no-tillage, direct drilling of the seeds into the soil). The aggregates in the 53- to 106-μm and 106- to 250-μm categories seem to be most affected by fungal growth and proliferation, which is more prevalent in the no-tillage treatments as compared with bacteria-dominated conventional tillage systems (Beare et al., 1994a,b).

It is the interactions among physical, chemical, and biological agents in soils which are so fascinating, complex, and important to consider as we increase the intensity of management of terrestrial ecosystems, or alter their usage in response to increased human concerns about their use, and also strive for effective sustainability of them worldwide (Coleman et al., 1992).

SUMMARY

The physical properties of the soil are the production of continued interactions between soil biota and their abiotic milieu. Water, the "universal solvent," casts a strong influence on the biota since many are adapted to life in a saturated atmosphere. The interplay between liquid and gaseous phases of water, in turn, is largely determined by pore size. The arrangement of particles in soils (the porosphere) is an important determinant for the ecology of the soil fauna.

Soil formation, the product of climate, organisms, parent material, and topology, leads to various soil types. Profile development and soil

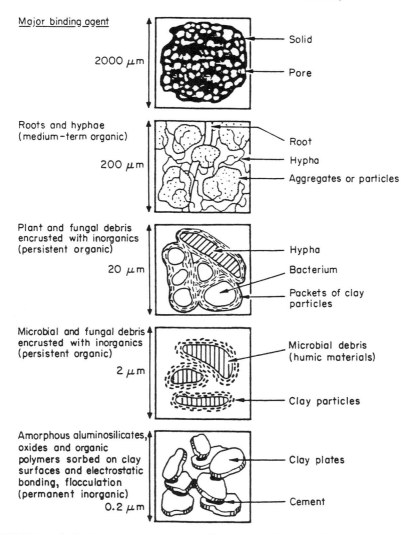

FIGURE 1.9 Soil microaggregates, across five orders of magnitude, beginning at the level of clay particles, through plant and fungal debris, up to a 2-mm-diameter soil crumb (from Tisdall and Oades, 1982).

texture are the product of interactions of these factors. The capabilities for nutrient retention, important for primary producers in all soils, are affected by both mineral content and soil organic matter, with organic matter usually having the higher number of exchange sites. The aggregate structure of soils is biologically mediated in many soil types.

2 | *Primary Production Processes in Soils*

INTRODUCTION

A. J. Lotka (1925), in his classic overview of ecological function, considered the system-level features of carbon gain, anabolism and system-level losses of this reduced carbon, or catabolism. This chapter is concerned with the primary sources of organic carbon inputs to soils, or system anabolism. These inputs have a major impact on N, P, and S dynamics, as will be shown in Chapter 5.

How can we best address the problems of measurement of primary production? Some ecological studies have declared that the accurate measurements of belowground inputs to ecosystems are virtually insurmountable and have assumed that belowground production equals that of production aboveground for total net primary production (NPP) (Fogel, 1985). This rule of thumb is clearly inadequate and often very wrong (Vogt *et al.*, 1986). Our objectives in this chapter are to address the processes and principles underlying primary production and to indicate where the "state of the art" is now, and is likely to be, over the next several years. A wide range of new techniques is now available. We anticipate that our understanding of belowground NPP will soon broaden.

THE PRIMARY PRODUCTION PROCESS

In the process of carbon reduction, a net accumulation of sugars, or their equivalents, exists in the organism's tissues. The costs of photosynthesis are extensively treated by plant physiologists and are out of the purview of this book. Other costs, related to movement of the photosynthates within the plant and allocation to symbiotic associates, are definitely relevant and will be considered further on.

Gross primary production minus plant respiration yields net primary production. Net primary production is the resultant of two principal processes: (a) increases in biomass, and (b) losses due to organic detritus production, which follows from, or is dependent on, the biomass production (Fogel, 1985). The detritus production includes leaves, branches, bark, inflorescences, seeds, and roots. Additional losses are traceable to exudation, volatilization, leaching, and herbivory (Cheng et al., 1993).

Measurement of aboveground components is at times tedious, but fairly complete in many studies (see reviews by Persson, 1980; Swank and Crossley, 1988). In contrast, measurement of belowground production processes has been fraught with errors and many difficulties. However, the total allocation of NPP belowground is often 50% or greater (Coleman, 1976; Harris et al., 1977; Fogel, 1985) (Table 2.1). A sizable portion of the total production is contributed by fine roots, which often have a high turnover rate, of weeks to months (Table 2.2), which may be closely linked to nitrogen availability on a seasonal basis (Nadelhoffer et al., 1985; Publicover and Vogt, 1993).

TABLE 2.1.   Annual Production ($Mg \cdot ha^{-1}$) of Fine Roots (<2 mm) and Root Production as Percentage of Total NPP in Different Ecosystems[a]

Ecosystem	Age (years)	% Contribution	Production
Coniferous forest			
Douglas fir	55	73	4.1–11.0
Loblolly pine	?	?	8.6
Scots pine	14	60	3.5
Deciduous forest			
Liriodendrun	80?	40	9.0
Oak–maple	80	?	5.4
Herbaceous			
Corn	< 1	25	1.2– 4.2
Soybean	< 1	25	0.6
Tallgrass prairie	?	50	5.1

[a]After Fogel (1985) and Coleman et al. (1976).

TABLE 2.2.   Annual Losses (Due to Consumption and Decomposition) of Fine Root Biomass in Different Forests[a]

Ecosystems	Loss (% total)
Deciduous forest	
European beech	80–92
Oak	52
Liriodendron	42
Walnut	90
Coniferous forest	
Douglas fir	40–47
Scots pine	66

[a]After Fogel (1985).

When comparing across ecosystems, one needs to be aware of marked differences in root morphology and distribution, i.e., root architecture (Fitter, 1985, 1991). Thus, wheat roots in a Kansas field are not markedly different in size, with primary and secondary laterals arising from root initials. In contrast, coniferous tree roots are often composed of long, supporting lateral roots and short roots which do the primary job of water and nutrient absorption. Ecologists often use a rather simple, pragmatic classification approach: (a) fine roots, less than 2 mm in diameter, and (b) structural roots, more than 2mm in diameter (Fogel, 1991).

METHODS OF SAMPLING

There are several methods for sampling roots, many of which have been reviewed by Böhm (1979). They may be generally classified into two principal approaches (Upchurch and Taylor, 1990): (a) destructive (sampling soil cores or monoliths) and (b) nondestructive, or observational, using rhizotrons or borescopes, termed minirhizotrons (Upchurch and Taylor, 1990; Cheng et al., 1990).

Destructive Techniques

The Harvest Method

This method involves taking samples, usually as soil cores, dry sorting the organic material, or rinsing it free by use of water or other flotation media, then sieving, sorting, and obtaining dry mass values. For sorting and categorizing roots, three factors need to be considered: root diameter, spatial distribution, and temporal distribution (Fogel,

1985). Much of the existing data have been derived from thousands of cores, washed, sorted, and analyzed by legions of weary graduate students and technicians. Some of these data have been truly informative and worth the effort. Other efforts, perhaps a majority of the published papers, have not. In the course of measuring root production by the harvest method, scientists often use what is known as the "peak-trough" calculation, in which the peaks and valleys of root standing crops through the course of a growing season as represented on a graph are successively added or subtracted about some general mean level. Unfortunately, there can be a fairly frequent occurrence of no net changes in root biomass, perhaps as often as 30% of the time in grasslands studies (Singh *et al.*, 1984), known as zero-sum years, which have no net production, as the increases in production are canceled out by those periods which show decreases. These problems were reviewed by Singh *et al.* (1984) (Fig. 2.1). They extensively analyzed a grassland root production data set, looking for effects of sample (replicate number) size and sampling frequency, coming to the conclusion that fairly frequently, perhaps in 3 years out of 10, one could expect to have no significant increments in growth, using the harvest method. In addition, they compared the amount of net primary production which one would expect from the peak-trough harvest method and multiple-year-based computer simulation model of root production and turnover. They found that the peak-trough method at times overestimated the "true" or simulated root production by as much as 150%, due to widely varying means, leading to spuriously high "production" values. The simulated production is not more real than the data of course, but they raised the question that perhaps the peak-trough method, as usually applied, often may lead to some significant overestimates of root production rates.

Isotope Dilution Method

Another approach, using the ^{14}C dilution technique, has been used to determine belowground biomass turnover in grasslands. Milchunas *et al.* (1985) performed a pulse labeling of plants for a few hours, then followed the time course of new ^{12}C label incorporated into both soluble and structural tissues of the root systems, a few weeks to months later. They then calculated the subsequent production on the assumption that any tissues lost would have a constant ratio of $^{14}C/^{12}C$ in the structural tissues. An additional step was to include ^{14}C incorporated into plant cell walls between sampling times 1 and 2. This greatly reduced the errors of the estimates (Table 2.3). Milchunas *et al.* (1985) found that the grass roots continued to mobilize additional amounts of ^{14}C from storage tissues in the grasses, but then made further measurements of the labeled plants to adequately account for the

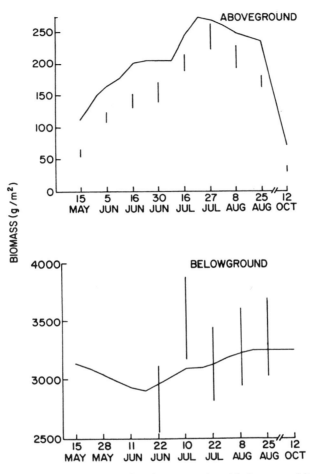

FIGURE 2.1 Comparison of values for aboveground and belowground biomass (g/m²) predicted by a simulation model with data collected in the field. The curves represent output of the model; vertical bars are means of field data plus and minus one standard error (from Singh *et al.*, 1984, with permission).

translocated ¹⁴C. Other researchers, notably Caldwell and Camp (1974), have used the isotope dilution technique with considerable success.

Root Ingrowth Technique

The root ingrowth technique (Steen, 1984, 1991) involves removing long cores of soil, sieving the soil free of roots, and then replacing the root-free soil into nylon tubular mesh bags with a mesh size of 5–7

TABLE 2.3. Comparison of Increments in the Belowground Biomass of Blue Grama and Wheat as Determined by Complete Harvest and $^{14}C/^{12}C$ Dilution Techniques[a]

| | Cell wall carbon | | | | Belowground production[b] | | | | |
| | | | | | Uncorrected | | | Corrected for ^{14}C incorporation | |
Time (days)	^{12}C (g)	$^{14}C(g \times 10^{-4})^c$	$R(\times 10^{-4})^d$	$R_c(\times 10^{-4})^e$	Harvest	$(R_1C_1/R_2C_2-1)B_1{}^f$	Error(%)[g]	$(R_1C_1/R_2C_2-1)B_1{}^f$	Error (%)[g]
Blue grama									
0.15	0.70	4.37	6.23						
5	0.90	22.56	25.07	10.58	3.52	-0.65	-118	3.48	-1
8	0.95	33.48	35.26	15.70	3.64	1.22	-66	3.56	-2
11	1.19	34.57	28.99	16.21	2.70	0.53	-80	2.73	1
25	1.40	39.67	28.25	18.60	2.15	0.78	-64	2.14	0
40	2.13	48.04	22.52						
Wheat									
0.15	0.53	1.82	3.46						
5	0.52	5.46	10.46	2.52	3.45	1.02	-70	3.53	2
8	0.72	6.91	9.60	3.19	2.96	1.25	-58	2.97	0
11	0.77	7.64	9.99	3.52	2.73	1.35	-50	2.68	-2
25	1.26	8.73	6.95	4.02	1.31	0.51	-56	1.33	2
40	2.02	9.83	4.87	4.53	-0.20	-0.35	75	-0.15	-25
62	2.17	10.19	4.70						

Note. Values are expressed on an ash-free dry weight basis.

[a]Modified from Milchunas et al. (1985), by permission of Kluwer Academic Publishers.

[b]Increment is grams of total root (structural plus labile) between that sampling time and the last sampling time.

[c]^{14}C = (DPM sample) (60sec/min) [Ci(3.7 × 10^{10} DPS)] (0.22442 g/Ci).

[d]$R = {}^{14}C/{}^{12}C$; $R_1 = {}^{14}C/{}^{12}C$ for that sampling time; and $R_2 = {}^{14}C/{}^{12}C$ for the last sampling time.

[e]$R_c = R_2$ corrected for the amount of ^{14}C incorporated into structural material between that sampling time and the last sampling time.

[f]C = % cell wall at time 1 or 2; B_1 = biomass (structural plus labile) at time 1.

[g](Dilution − harvest)/harvest × 100.

mm. The mesh bags are inserted by drawing them over a plastic tube, and the tube plus mesh bag on the outside is inserted into the hole in the field soil. The soil is tamped down in 5-cm increments as the tube is gradually withdrawn, leaving the mesh bag in position in the hole. Care must be taken to have a bulk density similar to that in the surrounding matrix. After the soil mesh bags are placed in their respective holes, roots are allowed to regrow into the bags. The bags are then recovered at various intervals, and the living and recent dead root biomass are measured (Hansson *et al.*, 1991; Steen, 1991). The principal assumptions are that growth into the root-free soil is the same as the root production would have been in the normal, undisturbed soil. Some concerns one might have about this technique are: was the bulk density of the soil in the mesh tubes identical to that in the surrounding soil? Also, were any significant soil aggregates broken in the soil-sieving process, which might alter the rates of root growth in the bags? Larger soil aggregates might be left intact if they do not contain any roots (Steen, 1991). What effect is caused by higher water contents in soil volumes without living roots? Advantages of the technique include being able to get a clear, more accurate measure of production of roots over discrete time intervals. Also, it is possible to obtain information on the decomposition of dead roots by placing fine-mesh (< 0.1 mm) cloth bags into the soil cylinders and determining the loss rates of dead roots simultaneously with the measurement of new root ingrowth over time (Titlyanova, 1987). However, for comments about live root and organic matter interactions, the reader is referred to Reid and Goss (1982) and Cheng and Coleman (1990). We will address some of these concerns further in the course of discussing decomposition processes in soils.

Nondestructive Techniques

Rhizotrons and minirhizotrons: A great resurgence of interest has been made in observational, nondestructive techniques for studying root-related processes. Several review volumes present detailed coverage, namely: Taylor (1987) and Box and Hammond (1990) among others. In essence, the rhizotron approach involves installing a large glass plate in an observation gallery, and then measuring the growth of roots against the glass over time (Fogel and Lussenhop, 1991) (Fig. 2.2). Using this technique, it is possible to follow a large part of a given root population visible through the glass over various time periods. The disadvantages are that the soil profile must be recreated and retamped to an equivalent bulk density or mass per unit volume of soil, which closely approximates that of the surrounding soil. It is also only a small fraction of an entire field or forest.

FIGURE 2.2   Soil biotron of the University of Michigan. (a) View of tunnel and above-ground laboratory from the south. (b) View from the west. Note white pine stump left after logging and burning in about 1917. (c) Interior of tunnel showing window bays covered with insulated shutters. (d) A close-up of glass; wire reinforced, 6 mm thick window pane. Note fungal rhizomorphs. The wire grid is about 2 × 2 cm (from Fogel and Lussenhop, 1991).

Minirhizotrons, on the other hand, have a smaller amount of surface area in one place, being tubular (5–7 cm diameter), and are placed, as are the rhizotrons, at a 20–25° angle from the vertical (Fig. 2.3). However, being light and readily handled, they enable extensive replication in any given plot, experimental treatment, or entire field site. Tubes may be of either glass or a durable plastic, such as polycarbonate. For example, Cheng *et al.* (1990) used 12 minirhizotron tubes in each replicate and two replicates per treatment (conventional tillage and no-tillage) in a study of sorghum root growth and turnover in a southeastern agroecosystem. Other studies have followed the dynamics of soil mesofauna, namely collembola, in fields under various crops in Michigan agroecosystems (Snider *et al.*, 1990). A number of precautions should be employed in the usage of minirhizotrons so as to avoid artifacts of placement. For example, total root biomasses

FIGURE 2.3   An auger jig system used to install angled minirhizotron tubes (from Mackie-Dawson and Atkinson, 1991).

can be underestimated in the top 7–10 cm if inadequate care is taken to shield the top of the minirhizotron tubes from transmitted light. Also, adequate soil/tube contact needs to be ensured by careful drilling and smoothing of the bored hole (preferably using a hydraulic coring apparatus), as noted by Box and Johnson (1987). If there is some open space between the outer tube surface and the soil, roots may respond as if this is a major soil crack, and preferentially grow along it (van Noordwijk *et al.*, 1993).

   To handle the large amounts of data and images obtained using minirhizotrons, it is necessary to use image analysis programs, such as those described by Smucker *et al.* (1987) and Hendrick and

Pregitzer (1992). With the advent of digital analysis techniques and image storage on CD-ROM, the literally millions of bits of information per soil/root image can be handled reasonably promptly and efficiently.

ADDITIONAL SOURCES OF PRIMARY PRODUCTION

An additional contribution to net primary production comes from algal populations in the surface few millimeters or on the soil surface itself. By measuring CO_2 fixation by cyanobacteria and algae on the surface of intact cores taken from an agroecosystem, Shimmel and Darley (1985) calculated that approximately 39 g of carbon was fixed per m² per year. This is a small proportion (5%) of total NPP for the study site, the conventional tillage agricultural system in Georgia (noted earlier), which averaged 800 g/m²/year aboveground NPP. The type of organic matter and the amount which may feed directly into detritivorous fauna could be of importance beyond the total production figures on an annual basis.

SYMBIOTIC ASSOCIATES OF ROOTS

From the earliest origins of a land flora, over 400 million years ago in the early Devonian, a structural/functional interaction has existed between plant roots and arbuscular mycorrhizae (AM) (Fig. 2.4a) as evidenced by the fossil record (Pirozynski and Malloch, 1975; Malloch *et al.*, 1980). Most families of terrestrial plants have mycorrhizal symbionts, with two families, the Cruciferae and Chenopodiaceae, being possible exceptions (Allen, 1991). The ectomycorrhizae (ECM) (Fig. 2.4b) are prevalent in several tree families, such as the Fagaceae (including the beeches and oaks) and the Pinaceae within the conifers. ECM arose relatively recently, only 160 million years ago, in the Cretaceous (St. John and Coleman, 1983).

After examining the structures of both principal types of mycorrhizae, we will consider information on carbon costs to the plant as well.

MYCORRHIZA STRUCTURE AND FUNCTION

Arbuscular mycorrhiza (AM), the so-called endomycorrhiza, are characterized by structures within root cells, called arbuscules, because they grow and ramify, tree-like, within the cell (Fig. 2.4a). They are members of the Phycomycete fungi. Most, but not all, AM (two exceptions are known in the family Endogonaceae) also have

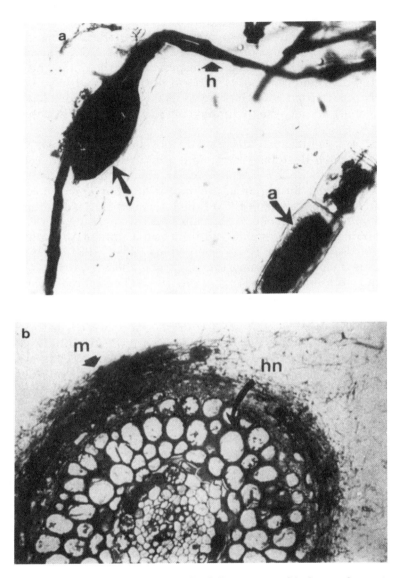

FIGURE 2.4   The internal structures of the different mycorrhizal types from ectomyc-orrhizae of *Quercus dumoa* (b) and arbuscular mycorrhizae of *Adenostoma sparsifolia* (a). Arbuscules (a), vesicles (v), internal hypae (h), mantle (m), and hartig net (hn) (from Allen, 1991).

storage structures known as vesicles, which store oil-rich products. AM send out hyphae for several centimeters (a maximum of 6–10) into the surrounding soil and are instrumental in facilitating nutrient uptake, particularly phosphate ions (Allen, 1991). AM are known only as obligate mutualists (i.e., the root provides carbon, and the mycorrhiza tap an enhanced pool of mineral nutrients) and have not been cultured apart from their host roots. AM are also avid colonizers of organic matter, and assist in enhancing soil structure (St. John *et al.*, 1983 a,b).

ECM are significantly different in physiology and ecology. These are principally Basidiomycetes, and proliferate between cells, not inside them as was the case for AM. An obvious morphological alteration occurs with formation of the mantle and hartig net (a combination of epidermal cells and ECM fungal tissues) on the exterior of the root (Fig. 2.4b). ECM send hyphae out literally several meters into the surrounding soil. The hyphae aid in nutrient uptake, including inorganic and some organic N + P compounds (Read, 1991). The hyphae constitute a significant proportion of carbon allocated to belowground NPP in coniferous forests (Vogt *et al.*, 1982). The reproductive structures of ECM are the often-observed mushrooms in oak or pine forests.

ECM will form resting stages, or sclerotia, which are cord-like bundles of hyphae that can persist for years. ECM, unlike AM, can often be cultured apart from their host plants. Some ECM may have considerable decomposing capabilities and can obtain a portion of their reduced carbon from decomposing substrates, i.e., leaf litter.

There are several other kinds of mycorrhiza, most notably Ericaceous mycorrhiza, which have some traits in common with ECM and AM. Ericaceous mycorrhiza are symbiotic with many Heathland plants; *Rhododendron* and *Kalmia* spp. are often infected with Ericaceous mycorrhiza (Dighton and Coleman, 1992). For information on these and other less common mycorrhiza, the reader is referred to Allen (1991, 1992) and Read (1991).

CARBON ALLOCATION IN THE ROOT/RHIZOSPHERE

Looking at the root/soil system as a whole, what is the totality of the resources involved, and how are these resources allocated under various conditions of stress and soil type?

Several reviewers (Coleman, 1976; Coleman *et al.*, 1983; Fogel, 1985, 1991; Martin and Kemp, 1986; Cheng *et al.*, 1993) have noted that from 20 to 50% more carbon enters the soil system from root exudates and exfoliates (sloughed cells and root hairs) than actually is present as fibrous roots, at the end of a growing season. This was determined in a series of experiments using ^{14}C as a radiotracer of the particulate and soluble carbon (Shamoot *et al.*, 1968; Barber and Martin, 1976). In fact, the mere change from a hydroponic medium to a sand medium was enough to double the amount of labile carbon as an input to the medium. This difference was attributed to the abrasion of roots against sand particles. In addition, the root/rhizosphere (Hiltner, 1904) microflora has the potential to act as a sizable carbon sink (Wang *et al.*, 1989; Helal and Sauerbeck, 1991), which can double the losses to soil as well. This is convincing proof that the combined belowground system—roots, microbes, soil, and fauna—is governed by source sink relationships, just as are intact plants (i.e., roots and shoots).

Extensive amounts and complexities of carbon compounds are elaborated in the rhizosphere (Rovira *et al.*, 1979; Kilbertus, 1980; Foster *et al.*, 1983; Foster, 1988; Lee and Foster, 1991; Cheng *et al.*, 1993). The boundary layer between root and soil, the so-called "mucigel" (Jenny and Grossenbacher, 1963), is jointly contributed by microbes and root surfaces.

CARBON ALLOCATION COSTS OF DEVELOPMENT AND MAINTENANCE OF SYMBIOTIC ASSOCIATIONS WITH ROOTS

It is difficult to measure apportioning of energy to roots because accurate measurement of belowground NPP entails a number of precautions as previously noted. Ever since the 1970s, there have been only a few estimates of the carbon costs which have been exacted by the fungal or rhizobial symbiont upon its root partner. The following are two examples of the sorts of measurements which have proven informative.

Pate *et al.* (1979) compared partitioning and utilization of assimilated carbon and nitrogen using nonnodulated, nitrate-fed, and nodulated, dinitrogen-fixing plants of white lupine, *Lupinus albus* L. Pate *et al.* (1979) calculated production and losses, and noted that not only were the nodulated root microflora more active, but there was also more new root growth under the stimulation of the nodule bacteria, which were acting as a greater root sink for translocated carbon.

Kucey and Paul (1982) measured two symbionts, an AM mycorrhiza and rhizobia in seedlings of faba beans, *Vicia faba* L. The bean seedlings were arranged experimentally as either mycorrhizal or nonmycorrhizal infected, and also as nodulated or nonnodulated bean plants, or four treatments in all. After inoculating or infecting the plants of choice, they then measured carbon dioxide fixation rates, translocation of ^{14}C labeled photosynthate to roots, and ^{15}N fixed by the various plants. They found a graded series of labeled ^{14}C translocated and/or evolved belowground as a function of infection complexity (Table 2.4). In addition, they obtained useful information on root and shoot weight, as well as rates of respiration.

The nodules of faba beans utilized 6% of the carbon fixed by singly infected (rhizobial) plants, but twice that amount, 12% of the C, by the doubly infected plants, i.e., both mycorrhiza and rhizobia symbionts (Table 2.5). Interestingly, the rates of CO_2 fixation increased significantly with biotic complexity, but changes in root and shoot biomass, although opposite to that of CO_2, were statistically insignificant.

Other studies, using real-time monitoring of ^{11}C under laboratory conditions (Wang *et al.*, 1989), found that mycorrhizal infection nearly doubled the "sink strength" of the roots for translocated photosynthate in studies of African Panicum grasses when compared with noninoculated control plants.

TABLE 2.4. ^{14}C Distribution in 5- to 6-Week-Old (8-hr Labeling Duration) Symbiotic and Nonsymbiotic Beans[a]

	Control	Mycorrhizal	Rhizobial	Mycorrhizal–Rhizobial
		Plant data		
CO_2 fixation rate[b]	6.79c[c]	6.96b	7.32b	9.34a
Shoot weight (g)	4.31a	4.40a	3.64a	3.59a
Root weight (g)	2.03a	1.65a	1.75a	1.64a
Nodule weight (g)	—	—	0.11	
Mycorrhizal infection (%)	—	58.6	—	54.8
		^{14}C distribution (%)		
Shoot biomass	54.6	52.20	46.8	42.0
Shoot respiration	1.7	1.0	2.0	1.1
Root biomass	20.7	20.2	25.0	16.8
Root respiration	23.0[d]	26.8[d]	24.6[d]	37.9[d]
Mycorrhizal biomass	—	ND	—	ND
Mycorrhizal respiration	—	ND	—	ND
Nodule biomass	—	—	1.61	2.24
Nodule respiration	—	—	ND	ND

[a]Reprinted from *Soil Biol. Biochem.* **14**, Kucey, R. M. N., and Paul, E.A., pp. 407–412, Copyright 1982, with kind permission from Elsevier Science Ltd, The Boulevard, Langford Lane, Kidlington OX5 1GB, UK.

[b](mg C g^{-1} shoot C hr^{-1}) calculated using shoot weights as measured at the end of the experiment.

[c]Means followed by the same letter do not differ ($P < 0.5$).

[d]Root plus symbiont respiration.

TABLE 2.5. N_2 Fixation by Mycorrhizal- and Nonmycorrhizal-Nodulated Beans (4 to 5 weeks old)[a]

	Nodule wt / root wt (mg g^{-1})	%N in shoot	N fixed (mg)	N fixed per unit nodule wt (mg g^{-1})
Rhizobial	87.7	3.81	0.78	16.2
Mycorrhizal–rhizobial	104.0[c]	4.34[b]	1.06[b]	15.8

[a]Reprinted from *Soil Biol. Biochem.* **14**, Kucey, R. M. N., and Paul, E.A., Carbon flow, photosynthesis, and N_2 fixation in mycorrhizal and nodulated faba beans (*Vicia faba* L.), pp. 407–412, Copyright 1982, with kind permission from Elsevier Science Ltd, The Boulevard, Langford Lane, Kidlington 0X5 1GB, UK.

[b]Significantly different (P ≤ 0.1%) from rhizobial treatment.

[c]Significantly different (P < 0.01%) from rhizobial treatment.

FUTURE RESEARCH ON ROOTS AND BELOWGROUND PROCESSES

As researchers and government agencies become ever more interested in, and concerned about, "sustainability" and long-term management of ecosystems, they will require much more information on system-level carbon allocation and energetics of these ecosystems.

The tools and analytical skills are at hand; it is now necessary to proceed with as much care in the assessment and measurement of belowground processes as has heretofore been given to aboveground processes.

SUMMARY

Primary production processes constitute the principal biochemical motive force for all subsequent activities of heterotrophs in soils. The inputs come in two directions: from aboveground onto the soil surface as litter, and belowground, as roots, which contribute exudates and exfoliated cells while the roots are alive, and then as root litter once the roots die.

A wide range of direct measurements of root production and turnover are now in use. These include various nondestructive techniques, including rhizotrons and minirhizotrons, and destructive techniques, including soil coring and isotopic labeling of roots, followed by destructive sampling at specified time intervals to determine dynamics, e.g., over an entire growing season.

Of equal importance to roots themselves are their generally more efficient physiological extensions, the root fungus mutualistic association, mycorrhiza. At a cost of 5–30% of the total photosynthate translocated belowground, mycorrhiza assist in obtaining inorganic nutrients and water over a much wider range of the soil volume than roots alone. This symbiotic association has a significant effect on other biota, namely microbes and fauna, which inhabit all soil systems.

3 | Secondary Production: Activities of Heterotrophic Organisms—Microbes

INTRODUCTION

We will now consider system-level catabolism (cata-bolos = breaking-down activity) or dissipation and transformation of energy. The transfer of energy from the primary producers into organisms farther along the food chain supports a wide range of heterotrophs. The production of new body tissues from primary production is called secondary production. If the plant food sources are living, the linkages are called a grazing food chain. Conversely, if the contributions from net primary production are dead, the sequence is termed a detrital food chain.

This difference in food chains has some impact on system function in that the feedback effects on the living tissues are direct rather than indirect, as in the detrital pathway. Soil food chains and webs are discussed further in Chapter 6.

The array of energy dissipators, or heterotrophs, in soil is incredibly diverse. The size range goes from <1 μm in length (bacteria) to the largest fossorial mammals, such as aardvarks or badgers, and giant earthworms, which reach several meters in length (Lee, 1985). Larger entities include ant and termite colonies, considered by some as a "superorganism" (Emerson, 1956). Larger yet are supercolonies of one organism of uniform genetic material, such as the extended mycelium of a fungus in Michigan which extended over more than 7 ha (Smith *et al.*, 1992).

All heterotrophs, of whatever size or volume, are involved in ingesting organic carbon and associated nutrients and assimilating them into carbohydrates, lipids, and proteins. Using a part for production of new body tissue, an extensive amount (40% or more) of the chemical bond energy is lost as metabolic heat and evolved CO_2. The initially synthesized plant carbohydrate (or its equivalent) is catabolized according to the general formula: $C_6 H_{12} O_6 + 6O_2 \rightarrow 6CO_2 + 6H_2O$; the more general formulation thus becomes $C_n H_{2n} O_n + nO_2 \rightarrow nCO_2 + nH_2O$. This is stoichiometrically the reverse of photosynthesis, which was discussed in Chapter 2.

COMPOUNDS BEING DECOMPOSED

Literally thousands of chemical and biochemical compounds are involved in catabolism. Viewed in an ecological context, however, they can be classified into two functional categories: (a) primary compounds, those which are directly derived from plant, microbial, or animal tissues, and (b) secondary compounds, those which are produced as a result of organic matter/mineral interactions, usually resulting in small or large chemical changes in chemical bonds or degree of aromaticity.

Both categories comprise a few major types (or groups) of compounds: soluble or labile versus relatively insoluble (in water), nonlabile, or resistant compounds. Compounds in the former category include organic acids, amino acids, and simple sugars. Compounds in the latter category include lignin, cellulose, cutins, and waxes. One should also consider biochemical vs biological bond types, as defined by McGill and Cole (1981). These reflect the differences between ester linkages, designated R-C-O-O-R which yield energy when broken, and the carbonyl C-N, C-P, or C-S bonds, which require energy when cleaved to yield nutrients for the microbes (Newman and Tate, 1980).

MICROBIAL ACTIVITIES IN RELATION TO CATABOLISM IN SOIL SYSTEMS

The principal "players" in the decomposition process are the microbial populations, i.e., the bacteria, fungi, and viruses. Bacteria and fungi are as biochemically diverse as they are diverse in phylogeny. Several genera of bacteria, being procaryotic, nave evolved the highly important biochemical trait of "fixing" (breaking the triple covalent bonds) of dinitrogen, making it available as ammonium for plant or microbial uptake (Postgate, 1987). This has important ramifications for nitrogen and phosphorus cycling and interactions with soil organic matter (Stewart and Cole, 1983; Stewart *et al.*, 1990).

Perhaps the greatest difference between bacteria and fungi is found in their mode of growth. Fungi have long strands (hyphae) which can grow into and explore many small microhabitats, secreting any of a considerable array of enzymes, decomposing material there, imbibing the decomposed subunits, and translocating them back in much or part of the hyphal network (Fig. 3.1). In contrast, bacteria are usually unicellular, or in clustered colonies, occupying discrete patches of soil measuring only a few cubic micrometers in volume. Bacteria depend on many episodic events for passive movement, such as rainfall, root growth, or ingestion by various soil fauna, to enable them to move about. When flagella are present, directed motility in the water film is also possible.

Viruses may play significant roles in the microbial ecologies of soil environments, as they can be a nonfaunal source of mortality, particularly for bacteria. Farrah and Bitton (1990) noted that lytic phages could act so as to restrict the growth of susceptible bacteria, and other phages could transmit genetic information between bacteria. The information on viral numbers and activities in soil in general is quite limited. Temperate phages (as distinct from virulent ones) in desert systems were inactivated on soil particles at acid pH (4.5–6). These phages had virtually no effect on populations of soil bacteria in Arizona soils, but persisted at low densities in their hosts (Pantastico-Caldas *et al.*, 1992). This contrasts markedly with the often-cited deleterious impacts of virulent phages on *Escherichia coli* in liquid cultures in the laboratory.

Unfortunately for the soil ecologist, the distribution and abundance of microorganisms are so patchy that it is very difficult to determine their mean abundances with accuracy without dealing with a very high variance about that mean when viewed on a macroscale. Part of this variation is due to the close "tracking" of organic matter "patches" by the microbes. There are aggregations of microbes around roots (the oft-cited "rhizosphere") (Lynch, 1990), around fecal pellets and other patches of organic matter (Foster, 1994), and in pore necks (Fig. 3.2) (Foster and Dormaar, 1991), and even concentrations of organisms in the mucus secretions which line the burrows of earthworms [the "drilosphere," as defined by Bouché (1975) and reviewed by Lee (1985)]. The phenomenon of "patches" is discussed more in Chapter 6.

Soil is an impressively heterogeneous matrix of minerals and organic matter. Ways in which this heterogeneity in organic matter

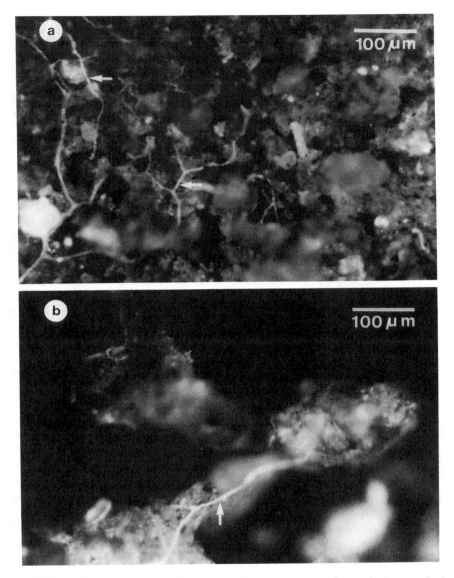

FIGURE 3.1 Extensive growth of fungal mycelium (arrow) was observed when crushed microaggregate (0.50 mm diameter) from native soil was stained with water-soluble aniline blue (a); smaller sized (0.10 mm diameter) aggregates from a crushed macroaggregate (1.0 mm diameter) were held together by fungal hyphae (b) (from Gupta, 1989).

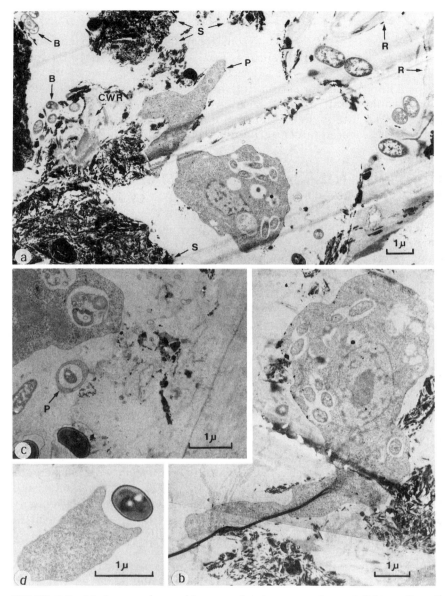

FIGURE 3.2 (a) An amoeba probing a soil microaggregate containing cell wall remnants (CWR) and a microcolony of bacteria (B). P, pseudopodium; R, root; S, soil minerals. Bar: 1 μm. (b) An amoeba with an elongated pseudopodium (P) reaching into a soil pore. The amoeba contains intact ingested bacteria in its food vacuoles. (c) An amoeba with partly digested bacteria in food vacuoles; note bacterium enclosed by a pseudopodium (P). (d) A pseudopodium associated with a gram-positive microorganism. Reprinted, with permission, from Foster, R. C., and Dormaar, J. F. (1991). Bacteria-grazing amoebae *in situ* in the rhizosphere. *Biol. Fertil. Soils* **11**, 83–87. Copyright 1991 Springer-Verlag.

and texture can influence microbial populations have been widely studied for more than a century. Textbooks, such as those by Swift *et al.* (1979) and Paul and Clark (1989), cover a number of methodological approaches for estimating microbial numbers and turnover in considerable detail. In this book, we present a few principal techniques and relate them to studies of nutrient immobilization and mineralization, covered later in Chapter 5 on decomposition processes.

TECHNIQUES FOR MEASURING MICROBIAL PRODUCTION AND TURNOVER

Techniques for measuring populations and biomass of microorganisms are either direct, by counting, or indirect, by inference from chemical and physical measurements. The following are a few of the more commonly used techniques for studies of microbial standing crops and activity in a community and ecosystem context.

Direct Measures of Numbers and Biomass

Total counts of microbes are made by preparation of soil (ca. 10 mg) suspensions spread in thin agar films on microscope slides (Jones and Mollison, 1948). The films are then stained, often with fluorescent dyes, and scanned. More recently, there have been improvements to the direct count technique, such as the membrane filtration technique, which enables one to quickly count fungal hyphae against the filter or a stained background. This approach is generally much faster and easier than the more laborious agar film technique. Other more classical techniques such as viable counts on nutrient-containing agar media are discussed by Parkinson *et al.* (1971) and Parkinson and Coleman (1991). The viable culture techniques usually recover only 1% or less of the total viable cells, so are useful only for comparative purposes, when one is focusing on a few readily culturable species of bacteria.

Other direct measures include sampling for extractable DNA (Torsvik *et al.*, 1990a,b; 1994), using techniques such as the polymerase chain reaction (PCR) to multiply substantially the DNA amounts and determine the identities of the organisms of interest. A more specific approach to microbial community analysis uses signature lipid biomarkers (SLB). This technique, pioneered by Dr. David White and colleagues at the University of Tennessee (Tunlid and White, 1992), measures ester-linked polar lipid fatty acids and steroids to determine microbial biomass and community structure.

Further comments on these techniques are given in Paul and Clark (1989).

The just mentioned procedures are primarily used in determining bacterial community structure. For fungi, several studies have made use of the fact that ergosterols are specific to fungi and that the amounts of ergosterols can be quantified to determine the amount of fungal tissues present in soils (Newell and Fallon, 1991; Eash et al. 1994).

Indirect Measures of Biomass

Chemical Methods

Jenkinson (1966) revived an earlier suggestion of Störmer (1908) that a flush of CO_2, evolved after fumigation, was due to the decomposition of organisms killed during fumigation by the surviving microorganisms remaining after fumigation. This relates to the extensive work done on "partial sterilization" of soils in Great Britain and elsewhere (Russell and Hutchinson, 1909; Powlson, 1975) under the misguided assumption that most soil microorganisms were somehow deleterious to subsequent plant growth, particularly in agricultural fields. Many of the soil heterotrophs are now considered generally beneficial, particularly when viewed in a whole-system nutrient cycling context.

The Chloroform Fumigation and Incubation Technique (CFI)

Using a fumigant such as chloroform and incubating the soil for 10 or 20 days, the size of the flush of CO_2 output can be related to the size of the microbial biomass by the expression: $B = F/k_c$; where B is the soil biomass C (in µg C^*g^{-1} soil); F is CO_2 minus C evolved by fumigated soil minus CO_2 evolved by unfumigated soil over the same time period; and k_c is the fraction of biomass mineralized to CO_2 during the incubation (Jenkinson and Powlson, 1976). The k_c value, calculated from a range of microorganisms in controlled experiments, is assigned a general value of 0.45 (Jenkinson, 1988).

Five assumptions are made in calculating biomass from the flush of CO_2 output: (1) The C in dead organisms is mineralized to CO_2 more rapidly than that in living organisms. (2) The kill is substantially complete. (3) Biomass dying in the unfumigated control soil is negligible when compared to that killed by fumigation. (4) The fraction of killed biomass C mineralized (k_c) is the *same* in different soils. There has been considerable discussion of how to calculate k_c. A commonly used value of 0.41 or 41% is used for evolution of CO_2, as distinguished from 0.6 (or 0.59) for production of new tissue. (5) Fumigation has *no* effect on the soil other than the killing of biomass—a *key* assumption. Experiments showed that similar amounts of CO_2 were evolved after either CFI or gamma irradiation treatments. Both treatments were considered to have affected the same labile microbial fraction of the soil organic matter (Jenkinson, 1966).

Jenkinson and Powlson (1976) relied on laboratory measurements of microbial cells added to soil. Voroney and Paul (1984) extended this work to include labile nitrogen and measured both k_c and k_n (fraction of biomass N mineralized to inorganic N). They took soil from southwestern Saskatchewan, incubated it for 1 month, and then amended it with ^{14}C-labeled glucose (1.58 µg C^*g^{-1} soil) and also ^{15}N atom % excess of 4.43, applied with a chromatographic fine-mist sprayer. They followed trends in assimilation of ^{14}C-labeled glucose, with and without additions of glucose and nitrate (Fig. 3.3) and mineral N extracted during incubation of glucose plus nitrate and nitrate-amended soils (Fig. 3.4). A more recent review of usage of ^{14}C to measure microbial biomass and turnover is given by Voroney *et al.* (1991), with step-by-step procedures for this research. They introduced carbon by labeling plants via photosynthetic pathways, then followed the carbon into the microbial biomass via root exudates and turnover, and in turn into the soil organic matter. A wide range of soils have been compared for biomass C calculated from biovolume (the measured volume of the cell), from the CFI method, and a ratio of biomass C from biovolume to biomass C from CFI (Powlson, 1994) (Table 3.1). These values range from 0.86 to 1.25 from arable lands and 6.47 from deciduous woodland. Forest soils, including those with low pH, have proven more difficult to analyze for microbial biomass and are considered next.

Chloroform Fumigation and Extraction Procedure (CFE)

Vance *et al.* (1987) noted that low pH soils, particularly those in the range below pH 5.0, including many forest soils, were not well characterized for microbial biomass using the CFI procedure. They modified the CFI procedure (Jenkinson and Powlson, 1976) as follows: soil samples are fumigated with chloroform for 48 hr, the fumigated and nonfumigated control samples are extracted with 0.5 M K_2SO_4, and the resulting organic extracts are measured for carbon, nitrogen, and other elements. This is denoted the chloroform fumigation and extraction procedure. For soils with pH values <4, the k_c values are usually lower, from 0.2 to 0.35 (Jenkinson, 1988). The CFE method has proven quite successful and enables one to obtain microbial

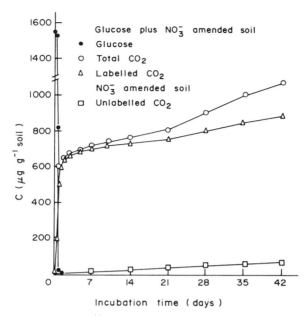

FIGURE 3.3   Assimilation of ^{14}C-labeled glucose and cumulative CO_2 evolved during incubation of glucose plus NO_3^- amended and NO_3^- amended soils. (Reprinted from "Soil Biology and Biochemistry," volume 16, Voroney, R. P. and Paul, E. A. Determination of K_c and K_n *in situ* for calibration of the chloroform fumigation–incubation method, pp. 9–14, Copyright 1984, with kind permission from Elsevier Science Ltd, The Boulevard, Langford Lane, Kidlington 0X5 1GB, UK.

biomass values for carbon, nitrogen, phosphorus (Hedley and Stewart, 1982), and sulfur (Gupta and Germida, 1988).

A few authors have expressed concern about the extent of faunal contributions to the fumigation "flush." Protozoan biomass may be a significant contributor in some soils (Ingham and Horton, 1987), but usually constitutes less than 2% of total microbial carbon.

Some general comments on the methodology of the microbial biomass method are necessary. The CFI and CFE methods should be employed within the context or intent of the methods originally described. Because they are bioassays, and not general chemical assays, they are not as robust as the latter. They can be misused, particularly if a great deal of organic matter substrate, waterlogging, or very low pH conditions are encountered (Powlson, 1994). However, the microbial biomass values are useful in the development and exercising of simulation models of labile carbon and nutrient turnover in a wide range of ecosystems (e.g., Parton *et al.*, 1987; 1989a,b; Jenkinson

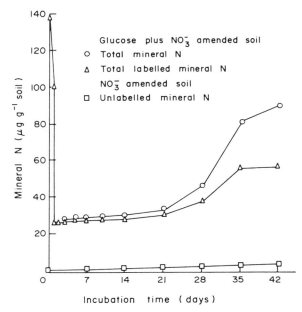

FIGURE 3.4   Mineral N extracted by 0.5 M K_2SO_4 during incubation of glucose plus NO_3^- amended and NO_3^- amended soils (Reprinted from "Soil Biology and Biochemistry," volume 16, Voroney, R. P. and Paul, E. A. Determination of K_c and K_n *in situ* for calibration of the chloroform fumigation–incubation method, pp. 9–14, Copyright 1984, with kind permission from Elsevier Science Ltd, The Boulevard, Langford Lane, Kidlington 0X5 1GB, UK.

and Parry, 1989). For a more general study of biochemical methods to estimate microbial biomass, the reader is referred to Ross and Sparling (1993).

Physiological Methods

Additional methods for measuring microbial biomass include the substrate-induced respiration (SIR) technique, first developed by Anderson and Domsch (1978). The SIR technique involves adding a substrate such as glucose to soil and measuring the respiration resulting from the stimulated metabolic activity in the experimental versus control treatments which received no carbon substrate. It is possible to measure the relative contributions of bacteria and fungi by using inhibitors, e.g., cycloheximide to inhibit fungal activity or streptomycin to inhibit bacterial activity. The assumption is that only bacterial activity is measured when fungi are inhibited and vice versa. The technique requires some care, as texture may affect the apparent

TABLE 3.1.   Comparison of Biomass Carbon as Calculated from Direct Microscopy and the Fumigation–Incubation (FI) Method[a]

Soil	Organic C (%)	pH	Biomass C calculated from biovolume[b] ($\mu g\ C/g$ soil)	Biomass C calculated from FI method[c] ($\mu g\ C/g$ soil)	Ratio of biomass C from biovolume to biomass C from FI
Arable[d]	2.81	7.6	550	547	1.01
Arable[d]	0.93	8.0	190	220	0.86
Deciduous woodland[d]	4.30	7.5	1540	1231	1.25
Arable[d]	2.73	6.4	390	360	1.08
Grassland[d]	9.91	6.3	3200	3711	0.86
Deciduous woodland[d]	2.95	3.9	330	51	6.47
Secondary rainforest[e]	1.46	7.1	430	540	0.80
Cleared forest[e, f]	1.23	6.2	260	282	0.92

[a]From Powlson (1994). Adapted from *Soil Biol. Biochem.* **8**, Jenkinson, D. S., Powlson, D. S., and Wedderburn, F. W. M., pp. 189–202, Copyright 1976, with kind permission from Elsevier Science Ltd, Pergamon Imprint, The Boulevard, Langford Lane, Kidlington 0X5 1GB, UK.
[b]See Jenkinson et al. (1976) for method of calculation.
[c]Calculated using Kc-0.45, not 0.5 as in the original paper.
[d]Temperate soils from the United Kingdom.
[e]Subhumid tropics, Nigeria.
[f]Arable cropping for 2 years after clearing secondary forest, Nigeria.

"resistance" to biocides. Further details of the technique are given by Beare *et al.* (1990; 1991); Insam (1990); Kjøller and Struwe (1994); and Alphei *et al.* (1995).

Additional Physiological Methods of Measuring Microbial Activity

A large body of literature deals with the indirectly measured signs of metabolic activity, namely CO_2 output or oxygen uptake. The ratio of the two gases, in terms of either uptake or output, is very informative about the principal sources of carbonaceous compounds being metabolized. The ratio of CO_2 evolved to oxygen taken up, or RQ, is lowest for carbohydrates, intermediate for proteins, and highest when lipids are the principal substrate being metabolized (Battley, 1987). In several studies the microbial respiration per unit microbial biomass ($qCO_2 = \mu g\ CO_2\text{-}C/mg\ C_{mic}/hr$; Anderson and Domsch, 1978; Insam and Domsch, 1988; Anderson and Domsch, 1993; Anderson, 1994) was measured and found useful as an indicator of the overall metabolic status of a given microbial community. Additional metabolic quotients have been used to study influences of climate and temperature, soil management, impact of heavy metals, and soil animals in ecosystems, notably the microbial carbon/organic carbon ratio expressed as a percentage of microbial carbon to total organic carbon, or C_{mic}/C_{org} (Table 3.2) (Anderson, 1994; Joergensen *et al.*, 1995). This follows from the assumption that terrestrial ecosystems in a near steady state are characterized by a constant flow of nutrients and energy, into and out of the ecosystem on a yearly basis, and entering and leaving the microbial biomass pool as well (Fig. 3.5; Anderson, 1994). All of the foregoing is based on aerobic conditions. The extent of anaerobicity can be important at certain times and needs to be carefully measured.

Enzyme Assays and Measures of Biological Activities in Soils

Numerous soil biologists/ecologists have attempted to use enzyme assays to measure soil biological activity (Coleman and Sasson, 1980; Nannipieri, 1994). Oxidoreductases, transferases, and hydrolases have been most studied. These assays have been considered of questionable value, mostly because of misapplication of the techniques and misinterpretation of the resulting data. The principal objection to soil enzyme assays is that the activities are substrate

TABLE 3.2.   Examples of Studies in Soil Microbiology in which Metabolic Quotients Have Been Applied[a]

Field of Study	*Metabolic quotient[b]*
Maintenance carbon requirement	m, qCO_2
Carbon turnover	qCO_2, μ, K_mGLUCOSE. Y, m, qD, C_{mic}/C_{org}
Soil management	qCO_2, qD, V_{max}, C_{mic}/C_{org}
Impact of climate and temperature	qCO_2, C_{mic}/C_{org}, qD
Impact of soil texture and soil compaction	qCO_2, qD
Impact of heavy metals	qCO_2
Ecosystems, ecosystem theory	qCO_2, qD, C_{mic}/C_{org}
Impact of soil animals	qCO_2

[a]Modified from Anderson (1994). See Anderson (1994) for specific references pertaining to usage of particular metabolic quotients.

[b]m, Maintenance coefficient; qCO_2, metabolic quotient or specific respiration rate; μ, specific growth rate; K_m, Michaelis–Menten constant; V_{max}, maximum specific uptake rate; Y, growth yield; qD, specific death rate; C_{mic}/C_{org}, microbial carbon/organic carbon ratio expressed as a percentage of microbial carbon to total organic carbon.

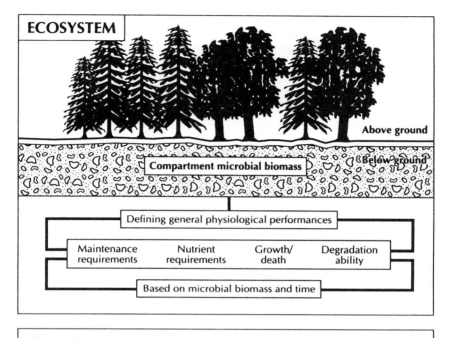

The prevailing assumption is that terrestrial ecosystems in a quasi–steady state are characterised by a constant flow of nutrients and energy, entering and leaving the system on a yearly basis, as well as entering and leaving the microbial biomass compartment. Microbial biomass communities adapt to the flow rate specific to the system.

C_{mic}/C_{org} ratio = percentage of microbial carbon to total soil carbon

Metabolic quotient for CO_2 (qCO_2) = μg CO_2-C/mg C_{mic}/h

FIGURE 3.5 Working hypothesis for the application of metabolic quotients in ecosystem development at the synecological level (from Anderson, 1994).

specific and are hence related to specific reactions and not necessarily reflecting organismal activities (Nannipieri *et al.*, 1990; Nannipieri, 1994). There is a continuing need to effectively distinguish between intracellular enzyme activity, reflecting ongoing microbial activity, and extracellular or "abiotic" (Skujins, 1967) activity, which reflects previous organismal activity, with the organisms themselves no longer existent (Nannipieri, 1994).

It should be noted that enzymes related to particular target substrates, such as lignocellulases in leaf litter, may be relatively good predictors of mass loss. After early stages of mass loss, due to leaching and mineralization, the "middle stage" is often strongly correlated with enzyme activity. In the final stages, ca. <25% of initial mass remaining, the accumulation of humic condensates depresses microbial activity, stabilizing the remaining material (Fig. 3.6) (Sinsabaugh *et al.*, 1994).

Direct Methods of Determining Biological Activity

Direct measurements of the activity of soil microorganisms have been a goal of soil biologists for a long time (Newman and Norman, 1943). This reflects the basic thermodynamic fact that as organisms undergo metabolic activity, they emit heat from the enthalpy of reactions occurring in net catabolism (Battley, 1987). With the continuing trend toward miniaturization of circuitry and much better, more sensitive thermocouples, it is now possible to obtain direct measures of metabolic activities of organisms in small samples of soils with only a few milligrams of biomass (Sparling, 1981; Battley, 1987). Flow microcalorimeters are now available which allow for the simultaneous measurements of CO_2 and NO_2 production in soils (Albers *et al.*, 1995).

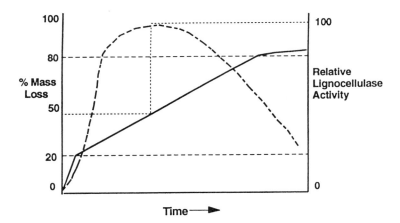

FIGURE 3.6   Idealized plot of lignocellulase activity (heavy dashed line) in relation to litter mass loss (solid line) through time. Lignocellulase activity traces a "bell-shaped" pattern over the course of litter decomposition, peaking (in this example) when the cumulative mass loss reaches 45%. The dashed horizontal lines at 20 and 80% mass loss highlight breakpoints in the mass loss curve. During the early stages of litter decomposition, rapid mass loss is often largely attributable to leaching and mineralization of soluble litter constituents. During the middle stage, lignocellulose degradation predominates. During the late stages, the accumulation of humic condensates depresses microbial activity, stabilizing the remaining material (from Sinsabaugh *et al.*, 1994).

SOIL STERILIZATION AND PARTIAL STERILIZATION TECHNIQUES

A number of the techniques mentioned earlier involve drastic perturbations to soils, such as fumigation, for the purpose of determining numbers or biomasses of organisms residing within them. Huhta *et al.* (1989) examined the influence of microwave radiation on soil processes, noting that it seems to have a less drastic impact compared to autoclaving, gamma irradiation, or chloroform fumigation. Microwaving is particularly useful for removing various mesofaunal groups, leaving the microbial communities reasonably intact under moderate thermal energy inputs of 380 W for 3 min (Huhta *et al.*, 1989; Wright *et al.*, 1989). Unfortunately, some unwanted side effects were introduced with microwaving, principally a decreased water-holding capacity. Monz *et al.* (1991) found that microwaving was of limited use in their agricultural soils. We discuss methods to manipulate fauna in Chapter 4.

CONCEPTUAL MODELS OF MICROBES IN SOIL SYSTEMS

As was noted in the discussion of primary production processes (Chapter 2), there are "hot spots" of activity, particularly of microbes in relationship to root surfaces and rhizospheres. When viewed as a transect through the rhizosphere, e.g., 2 mm from the root surface or less, there are arrays of rapidly growing bacteria and fungi, which have been called "fast" flora (Trofymow and Coleman, 1982) (Fig. 3.7). Moving up the root toward the shoot, into older regions, root hairs, and then root cortical cells, which may be sloughing off into the surrounding soil, are found. There are accompanying microbial and root grazers, such as protozoa and nematodes, which are discussed in detail in Chapter 4. Out in the bulk soil, away from the rhizosphere (>4 mm from the root surface), occur some of the slower-growing or "slow" bacteria and fungi, organic matter fragments, and some of the hyphae of either AM or ectotrophic mycorrhiza.

SUMMARY

The process of consumption and decomposition is considered ecologically as system-level catabolism. The primary agents of decomposition are bacteria and fungi, often considered as "microbial biomass."

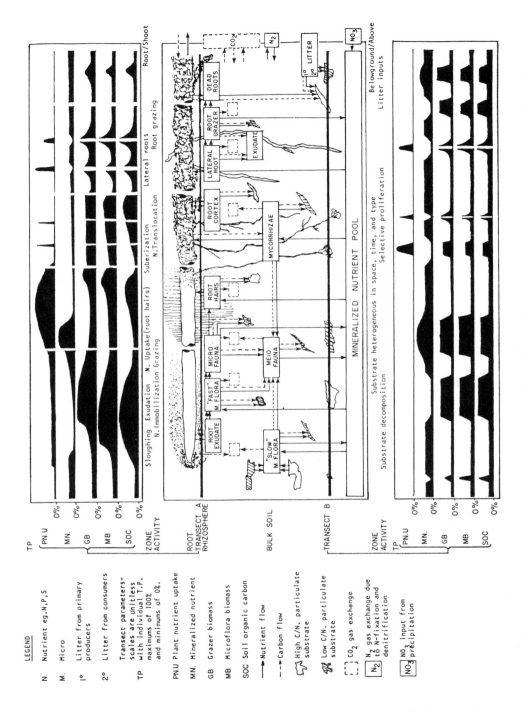

FIGURE 3.7 Conceptual diagram of a root/rhizosphere/soil system (from Trofymow and Coleman, 1982).

Microbial production and turnover are determined in a number of indirect ways, both chemical and physiological, as well as in a direct fashion, using high-magnification microscopy or via energetics approaches such as calorimetry.

The microbial biomass, while relatively small (ca. 200–400 g*m^{-2} in the surface 15 cm), has a rapid turnover time and serves as a principal food source for microbivorous fauna and is also the source of labile nutrients available for plant roots and other microbes. Hence the microbial community is indeed the "eye of the needle" through which virtually all of the decomposition carbon and nutrients must pass.

4 | *Secondary Production: Activities of Heterotrophic Organisms—The Soil Fauna*

INTRODUCTION

Animals, the other group of major heterotrophs in soil systems, exist in elaborate food webs containing several trophic levels. Some soil animals are true herbivores as they feed directly on roots of living plants, but most subsist on dead plant matter, microbes associated with it, or a combination of the two. Still others are carnivores, parasites, or top predators. Actual heterotrophic production by the soil fauna is only poorly known. Analyses of food webs in the soil have emphasized numbers of the various organisms and their trophic resources. Estimates of biomass of soil animals are less common, and knowledge of the rates of energy or material transfer in food webs is fragmentary (Moore and de Ruiter, 1991). Analysis of the structure of these food webs reveals complex structures with many "missing links" poorly described (Walter *et al.*, 1991).

Animal members of the soil biota are numerous and diverse. The array of species is very large, including representatives of all terrestrial Phyla. Many groups of species are poorly understood taxonomically and details of their biology are unknown. The soil fauna encompasses a rich pool of species. Protection of biodiversity in ecosystems must include soil species. Communities of soil fauna offer opportunities for studies of phenomena such as species interactions, resource utilization, or temporal and spacial distributions.

Soil ecologists cannot hope to become experts in all animal groups. When research focuses at the level of the soil ecosystem, two things are required: the cooperation of zoologists and the lumping of animals into functional groups. These groups are often taxonomic, but species with similar biologies are grouped together for purposes of integration (Coleman *et al.*, 1983; 1993; Hendrix *et al.*, 1986).

The soil fauna also may be characterized by the degree of presence in the soil (Fig. 4.1) or microhabitat utilization by different life forms. Transient species are exemplified by the ladybird beetle, which hibernates in the soil, but otherwise lives in the plant stratum of the garden. Gnats represent temporary residents of the soil since the adult stages live above ground. Their eggs are laid in the soil and their larvae feed on decomposing organic debris. In some soil situations dipteran (fly) larvae are important scavengers. Cutworms also are temporary soil residents, whose larvae feed on seedlings by night and hide by day. Periodic residents spend their life histories below ground, with adults such as the velvet mites emerging perhaps to reproduce. Collembolans represent permanent soil residents (Fig. 4.1). Even permanent residents of the soil may be adapted to life at various depths in the soil.

The morphology of collembolans displays their adaptations for life in different soil strata (see Fig. 4.10). Species that dwell on the soil

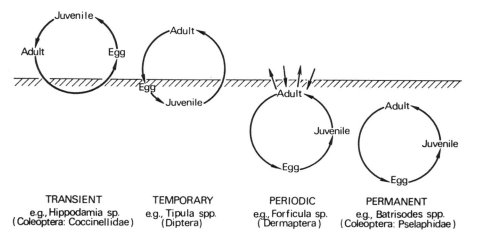

TRANSIENT	TEMPORARY	PERIODIC	PERMANENT
e.g., *Hippodamia* sp.	e.g., *Tipula* spp.	e.g., *Forficula* sp.	e.g., *Batrisodes* spp.
(Coleoptera: Coccinellidae)	(Diptera)	(Dermaptera)	(Coleoptera: Pselaphidae)

FIGURE 4.1   Categories of soil animals defined according to degree of presence in soil, as illustrated by some insect groups (Wallwork, 1970).

surface or in the litter layer may be large, pigmented, and equipped with long antennae and a well-developed jumping apparatus (furcula). Within the mineral soil collembolans tend to be smaller with unpigmented, elongate bodies and much reduced furculae—there is no place to jump to.

Numerous researchers have marveled at the many and varied body plans and size differences of the soil fauna. A generalized classification by length (Fig. 4.2) illustrates a commonly used device for separating the soil fauna into size classes: microfauna, mesofauna, macrofauna, and megafauna. This classification encompasses the range from smallest to largest, i.e., from ca. 1–2 µm of the microflagellates to several meters for giant Australian earthworms.

Body width of the fauna is related to their microhabitats (Fig. 4.3). The microfauna (protozoa, small nematodes) inhabit water films. The mesofauna inhabit air-filled pore spaces and are largely restricted to

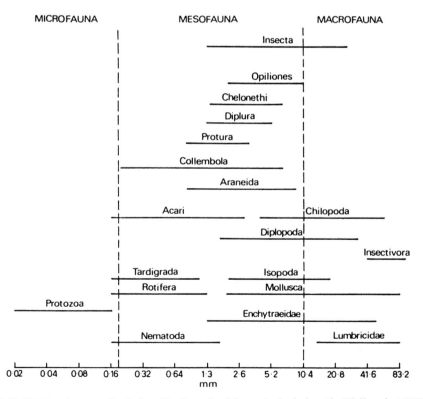

FIGURE 4.2   A generalized classification of soil fauna by body length (Wallwork, 1970).

existing ones. The macrofauna, in contrast, have the ability to create their own spaces through their burrowing activities, and like the megafauna, can have large influences on gross soil structure (Lee, 1985). Methods for studying these faunal groups are in large part size dependent. Methods for studying the microfauna rely mainly on techniques used for microbiology. Mesofauna require microscopic techniques for study and specialized extraction procedures for collection. The macrofauna may be sampled as field collections, often by hand-sorting, and populations of individuals are usually measured.

There is, of course, considerable gradation in the classification based on body width. The smaller mesofauna exhibit characteristics of the microfauna and so forth. Nevertheless, the classification continues to have considerable utility.

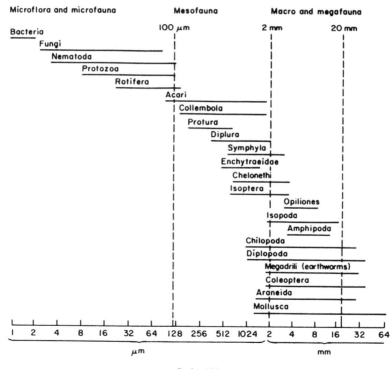

FIGURE 4.3   Size classification of organisms in decomposer food webs by body width (Swift *et al.*, 1979).

MICROFAUNA

The free-living protozoa of litter and soils belong to two Phyla: the Sarcomastigophora and the Ciliophora (Levine *et al.*, 1980). For practical purposes, we consider them in four ecological groups: the flagellates, naked amoebae, testacea, and ciliates (Lousier and Bamforth, 1990). A general comparison of body plans is given in Fig. 4.4, showing representatives of the four major types. After a brief overview of the groups, we will consider aspects of their enumeration and identification.

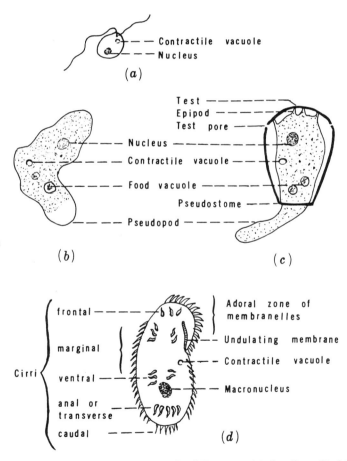

FIGURE 4.4   Morphology of four types of soil Protozoa: (a) flagellate (*Bodo*), (b) naked amoeba (*Naegleria*), (c) testacean (*Hyalosphenia*), and (d) ciliate (*Oxytricha*) (from Lousier and Bamforth, 1990).

Flagellates

Named for their one or more flagella, or whip-like propulsive organs, flagellates are among the more numerous and active of the protozoa. They play a significant role in nutrient turnover by their often intensive feeding activities, with bacteria as their principal prey items (Zwart and Darbyshire, 1991; Kuikman and Van Veen, 1989). Numbers have varied from 100 per g in desert soils to more than 10^5 per g in forest soils (Bamforth, 1980).

Naked Amoebae

Among the more voracious of the soil protozoa are naked amoebae. These organisms are very numerous and active in a wide range of agricultural, grassland, and forested soils (Elliott and Coleman, 1977; Clarholm, 1981, 1985; Gupta and Germida, 1989). The dominant mode of feeding for the amoebae, as for the larger forms such as Ciliates, is phagotrophic (engulfing), with bacteria, fungi, algae and other fine particulate organic matter being the majority of the ingested material (Bamforth, 1980; Bryant et al., 1982). The highly plastic mode of existence of the naked amoebae is impressive in terms of the ability which they have to explore very small cavities or pores in soil aggregates and feed on bacteria which would otherwise be considered inaccessible to predators (Foster and Dormaar, 1991).

Testate Amoebae

When compared with the naked amoebae, testate amoebae are often less numerous, except in moist, forested systems where they thrive. However, they are more easily censused by a range of direct filtration and staining procedures (Lousier and Parkinson, 1981). Detailed community production and biomass studies of testacea have been carried out in forested French sites by Coûteaux (1972, 1985) and in Canadian aspen forest lands (Lousier and Parkinson, 1984). For example, Lousier and Parkinson (1984) noted a mean annual biomass of 0.07 g dry wt/m² of aspen woodland soil, much smaller than the average annual mass for bacteria or fungi of 23 and 40 g, respectively. However, the testacean annual secondary production (new tissue per year) was 21 g dry wt/m², or essentially the entire average standing crop of the bacteria in that site.

Certain genera of testacea are also diagnostic of soil types. Foissner (1987) notes that pioneer soil scientists such as P. E. Mueller in the 1880s were able to differentiate between mull and mor forms of humus by the kinds of testacea found (Table 4.1). They used ratios of abundance of forms rather than exclusivity of presence or absence.

Ciliates

These protozoa have their own unusual life cycles and complex reproductive patterns, and tend to be restricted to very moist or seasonally moist habitats. Their numbers are lower than other groups, with a general range of 10 to 500 per g of litter/soil. Ciliates can be very active in entering soil cavities and pores and in exploiting bacterial food sources in them (Foissner, 1987). In common with other protozoa, ciliates have resistant or encysted forms from which they can emerge when conditions become favorable for growth and reproduction, with the presence of suitable food sources (Foissner, 1987). Ciliates, along with flagellates and naked and testate amoebae, can quickly reproduce asexually by fission. The flagellates, naked amoebae, and testacea can reproduce by syngamy, or fusion of two cells. For the ciliates, sexual reproduction occurs by conjugation, with the micronucleus undergoing meiosis in two individuals, and the two cells joining at the region of the cytostome and exchanging haploid "gametic" nuclei. Each cell then undergoes fission to produce individuals which are genetically different from the preconjugant parents (Lousier and Bamforth, 1990).

TABLE 4.1. Species Characteristic of the Testacean and Ciliate Communities in Mull and Mor Soils[a]

Type of humus	Testaceans (characteristic species)	Ratio of full and empty shells	Ciliophora (characteristic species)
Mull	Centropyxis plagiostoma C. constricta C. elongata Plagiopyxis minuta Geopyxella sylvicola Paraquadrula spp.	<1:2–5	Urosomioda agilis Urosoma spp. Hemisincirra filiformis Engelmanniella mobilis Grossglockneria hyalina Colpoda elliotti
Moder and Mor	Trigonopyxis arcula Plagiopyxis labiata Assulina spp. Corythion spp. Nebela spp.	>1:2–5	Frontonia depressa Bryometopus sphagni Dimacrocaryon amphileptoides

[a]From Foissner (1987).

As noted earlier for the testacea, some genera of ciliates are considered indicative of acid humus, whereas others are more typical of higher pH or "mild humus." Species characteristic of mull and mor soils can be found in Table 4.1.

Methods for Extracting and Counting Protozoa

There is an extensive literature on approaches to extracting and counting protozoa. For much of the 20th century, researchers have been heavily influenced by the papers of Cutler (1920, 1923), Cutler *et al.* (1923), and Singh (1946). These authors favored the culture technique, in which small quantities of soil or soil suspensions from dilution series are incubated in small wells, and are inoculated with a single species of bacteria as a food source. Based on the presence or absence in each well, the overall population density ("most probable number") can be calculated. Other scientists, notably Coûteaux (1972) and Foissner (1987), espouse the direct count approach, in which soil samples are examined in water to see what organisms are present in the subsample. The advantages of this approach are that it is possible to observe the organisms which are immediately present and not have to rely on the palatability of the bacterium used to inoculate the series of wells in the culture technique. The disadvantage of the direct count method, as noted by Foissner (1987), is that only 5–30 mg of soil, is usually employed so as not to be overwhelmed with total numbers. Unfortunately, this discriminates against some of the more rare forms of testaceans or ciliates which occur only infrequently, but may have a significant impact if they happen to be very large. Given rather limited research budgets, it is seldom possible to employ a small army of staff to scan literally hundreds of slides of soil from a single sample site.

An additional complication is the fact that the culture technique attempts to differentiate between active (trophozoite forms) and inactive (cystic) forms by the treatment of replicate samples with 2% hydrochloric acid overnight. The acid kills off the trophic forms and then, after washing in dilute NaCl, the counting continues. This assumes that all of the cysts will excyst after this drastic process; sometimes the assumption is met, but not always.

Distribution of Protozoa in Soil Profiles

Although protozoa are considered to be distributed principally in the upper few centimeters of a soil profile, they are also found at depth, over 200 m deep in groundwater environments (Sinclair and Ghiorse, 1989). Small (2- to 3- µm cell size) microflagellates were found to decrease 10-fold in numbers during movement through 1 m in a sandy matrix under a trickling filter facility (in dilute sewage) as compared to a 10-fold reduction in bacterial transport over a 10-m distance (Harvey *et al.*, 1995).

Impacts of Protozoa on Ecosystem Function

Several investigators have noted the obvious parallel between the protozoan–microbe interaction in water films in soil, on root surfaces, and in open water aquatic systems (Stout, 1963; Coleman, 1976, 1994a; Clarholm, 1994). The so-called "microbial loop" *sensu* Pomeroy (1974) has proven to be a powerful conceptual tool; rapidly feeding protozoa may consume several standing crops of bacteria in soil every year (Clarholm, 1985; Coleman, 1994b). Darbyshire and Greaves (1967) noted that this tendency is particularly marked in the rhizosphere, which provides a ready food source for microbial prey. This was demonstrated impressively for protozoa in arable fields (Cutler *et al.*, 1923) and, more recently, for bacteria, naked amoebae, and flagellates in the humus layer of a pine forest in Sweden after a rain (Clarholm, 1994). Bacteria and flagellates began increasing immediately after the rainfall event and rose to a peak in 2–3 days; naked amoebae rose more slowly and peaked at days 4–5 and then tracked the bacterial decrease downward, as did the flagellates (Fig. 4.5). Further information on protozoan feeding activities and their impacts on ecosystem function are given in Darbyshire (1994), Griffiths (1994), Zwart *et al.* (1994), and Pussard *et al.* (1994).

MESOFAUNA

Rotifera

Among the small fauna, these organisms are often found only when a significant proportion of water films exist in soils. They are usually considered to be aquatic organisms and may not be listed in major compendia of soil biota (Dindal, 1990); they are a genuine, albeit secondary component of the soil fauna (Wallwork, 1976). While sampling for nematodes in the surface layers of agricultural fields near La Selva in the Atlantic coastal forest of Costa Rica, one of us (DCC) found virtually no nematodes, but instead found large numbers of rotifers (tens of thousands per meter square), despite the soil being far from water saturation. The field was being maintained in a "bare fallow" regime with frequent weeding or denudation of vegetation to deliberately reduce organic inputs. However, there seemed to be ample

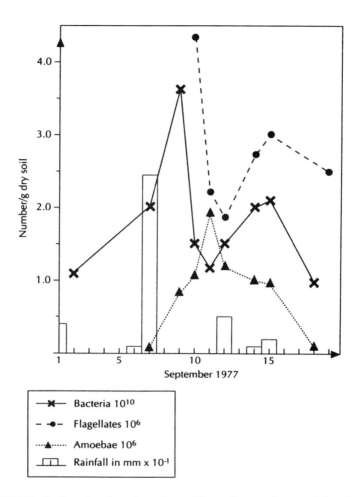

FIGURE 4.5   Daily estimates of numbers of bacteria, amoebae, and flagellates in the humus layer of a pine forest after rain (from Clarholm, 1994).

Cyanobacteria and perhaps other unicellular primary producers, which would have provided food for the rotifers. Some rotifers have been found in bagged leaf litter on forest floors in the southern Appalachian mountains.

Features of Body Plan and General Ecology

More than 90% of soil rotifers are in the order Bdelloidea, or worm-like rotifers. In these creeping forms, the suctorial rostral cilia and the adhesive disc are employed for locomotion (Donner, 1966). Rotifers also form cysts to endure times of stress or lack or resources. Additional life history features of interest include the construction of shells from a body secretion, which may have particles of debris and/or fecal material adhering to it. Some rotifers will use the empty shells of Testacea, the thecate amoebae. The Bdelloidea are vortex feeders, creating currents of water which conduct food particles to the mouth for ingestion (Wallwork, 1970). The importance of these organisms is largely unknown, although they may reach numbers exceeding 10^5 per m² in moist, organic soils (Wallwork, 1970).

For methods to extract and enumerate rotifers see methods used for extracting nematodes.

Nematoda

Nematodes, or roundworms, are among the most numerous of the multicellular organisms found in any ecosystem. As with the protozoa, they are primarily inhabiters of water films or water-filled pore spaces in soils. Nematodes have a very early phylogenic origin, but as with many other invertebrate groups, the fossil record is fragmentary. They are classified among the triploblastic pseudocoelomates (three body layers: ectoderm, mesoderm, and endoderm). In other words, nematodes have a body cavity for the gastrointestinal tract, but it is less well differentiated than that for the true coelomates, such as annelids and arthropods.

The overall body shape is cylindrical, tapering at the ends (Fig. 4.6). In general, nematode body plans are characterized by a "tube within a tube" (alimentary tract/the body wall). The alimentary tract, beginning at the anterior end, consists of a stoma or stylet, pharynx (or esophagus), intestine and rectum which opens externally at the anus. The reproductive structures are quite complex, as shown in Fig. 4.6. Some species are parthenogenetic, reproducing without sex. It is possible to view the internal structures of most nematodes because they have virtually transparent cuticles. The nematodes can be keyed out fairly readily to family and/or genus under a moderate magnification (ca. 100×) binocular microscope or in a Sedgwick-Rafter chamber on an inverted microscope (Wright, 1988), but species-specific characteristics must be determined under high magnification, using compound microscopes.

Nematode Feeding Habits

Nematodes feed on a wide range of foods. A general trophic grouping is bacterial feeders, fungal feeders, plant feeders, and predators and omnivores. For the purposes of our general overview, anterior (stomal or mouth) structures can be used to differentiate general

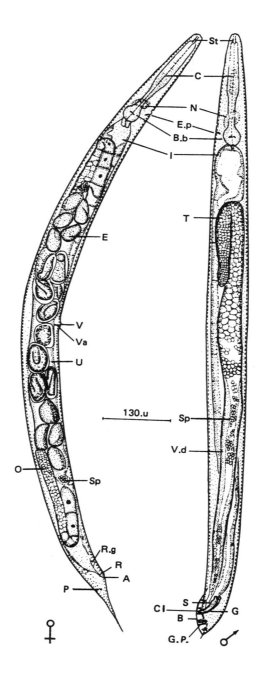

feeding or trophic groups (Fig. 4.7) (Yeates and Coleman, 1982; Yeates *et al.*, 1993). The feeding categories are a good introduction, but the feeding habits of many genera are either complex or poorly known. Thus immature forms of certain nematodes may be bacterial feeders and then become predators on other fauna once they have matured (Allen-Morley and Coleman, 1989). Some of the stylet-bearing nematodes, e.g., the family Neotylenchidae, may feed on roots, root hairs, and fungal hyphae (Yeates and Coleman, 1982).

Because of the wide range of feeding types and the fact that nematodes seem to reflect ages of the systems in which they occur [i.e., annual vs perennial crops (Neher *et al.*, 1995) or old fields and pastures and more mature forests], they have been used as indicators of overall ecological condition (Bongers, 1990; Ettema and Bongers, 1993; Freckman and Ettema, 1993). This is a growing area of research in soil ecology, one in which the intersection between community analysis and ecosystem function could prove to be quite fruitful. We discuss some of these concepts further in the section on decomposition and nutrient cycling.

Nematode Zones of Activity in Soil

As noted in Chapter 2, the rhizosphere is a zone of considerable metabolic activity for root-associated microbes. This extends also to the soil fauna, which may be concentrated in the rhizosphere. For example, Ingham *et al.* (1985) found up to 70% of the bacterial and fungal-feeding nematodes in the 4–5% of the total soil which was rhizosphere, namely the amount of soil 1–2 mm from the root surface (the rhizoplane). In comparison, Griffiths and Caul (1993) found that nematodes migrated to packets of decomposing grass residues, with considerable amounts of labile substrates therein, in pot experiments. They concluded that nematodes are seeking out these "hot spots" of concentrated organic matter and that protozoa, also monitored in the experiment, do not.

Nematode Extraction Techniques

Nematodes may be extracted by a variety of techniques, either active or passive in nature. For more accuracy in the determination of populations, the passive or flotation techniques are generally

FIGURE 4.6 Structures of a *Rhabditis* sp., a secernentean microbotrophic nematode of the order Rhabditida. ST, stoma; C, corpus area of the pharynx; N, nerve ring; E.p, excretory pore; B.b, basal bulb of the pharynx; I, intestine; T, testis; E, eggs; V, vulva; Va, vagina; U, uterus; O, ovary; SP, sperm; V.d, vas deferens; R.g, rectal glands; R, rectum; A, anus; S, spicules; G, gubernaculum; B, bursa; P, phasmids; G.P, genital papillae; CL, cloaca. *Courtesy of Proceedings of the Helminthological Society of Washington* (from Poinar, 1983).

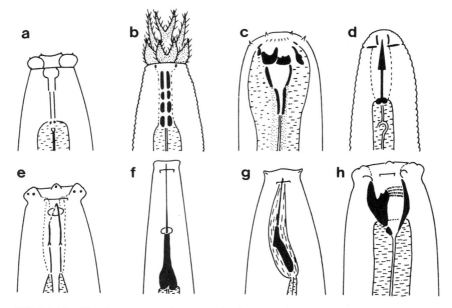

FIGURE 4.7   Head structures of a range of soil nematodes. (a) *Rhabditis* (bacterial feeding); (b) *Acrobeles* (bacterial feeding); (c) *Diplogaster* (bacterial feeding, predator); (d) tylenchid (plant feeding, fungal feeding, predator); (e) *Dorylaimus* (feeding poorly known, omnivore); (f) *Xiphinema* (plant feeding); (g) *Trichodorus* (plant feeding); (h) *Mononchus* (predator). [Reprinted from Yeates, G. W., and Coleman, D. C. (1982). Role of nematodes in decomposition. *In* "Nematodes in Soil Ecosystems" (D. W. Freckman, ed.), pp.55–80. Courtesy of the University of Texas Press, Austin.]

preferred. The principal advantage of the oldest, active method, namely the Baermann funnel method, is that it is simple, requiring no fancy equipment or electricity. It is based on the fact that nematodes in soils will move about in the wetted soil and fall into the funnel itself. Thus samples are placed on coarse tissue paper, on a coarse mesh screen, and then placed in the cone of a funnel and immersed in water. Once they crawl through the moist soil and filter paper, the nematodes fall down into the neck of the funnel. Because nematodes have only circular and not longitudinal muscles, they do not stay in suspension in the water and fall to the bottom of the funnel stem which was closed off with a screw clamp on a rubber hose. At the conclusion of the extraction (typically 48 hr), the nematodes in solution are drawn off into a tube and kept preserved for examination later. One drawback to the technique is that it allows dormant nematodes to become active and be extracted, so it may give a slightly inflated estimate of the true, "active" population at a given time. Other methods include filtration, or decanting and sieving, and flotation/centrifugation (Christie and Perry, 1951) to remove the nematodes from the soil suspension. For handling larger quantities of soil, up to 500 g, to recover large amounts of nematodes, various elutriation (extraction using streams of air bubbles in funnels) methods are employed. For details, see Górny and Grüm (1993) (Fig. 4.8).

Tardigrada

These interesting little micrometazoans [ranging from 50 µm, the smallest juvenile, to 1200 µm, the largest adult (Nelson and Higgins, 1990)] are also called "water bears" because of their microursine appearance. They were named "Il Tardigrado," literally slow-moving forms, by the famous Italian abbot and natural history professor, Lazzaro Spallanzani, in 1776 (Nelson and Higgins, 1990).

Tardigrades are considered to have affinities to the aschelminth complex (including nematodes) and the arthropoda. They are bilaterally symmetrical with four pairs of legs, equipped with claws on the distal end, of various sizes and forms (Fig. 4.9). The sizes and shapes of the claws are used in keying genera and species. Perhaps their greatest notoriety in recent times has come from the marked recuperative powers which they show after having been kept in a state of "suspended animation" for many years or even decades. These studies (Crowe, 1975; Crowe and Cooper, 1971) have found that Tardigrades recover well even after extreme environmental insults such as being plunged into liquid nitrogen. More generally, a series of five types of latency or virtual cessation of metabolism have been described: encystment, anoxybiosis, cryobiosis, osmobiosis, and anhydrobiosis (Crowe, 1975). All of these are subsumed under the more general term: "cryptobiosis" or hidden life. Being highly resistant, or resilient, to various environmental insults, these organisms exemplify a recurrent thread throughout biology in general, and soil biology in particular: the selective advantages to "waiting out" a spell of bad microclimate and being able to reactivate and become active in a given patch in the soil after years or decades have elapsed.

Tardigrades tend to occur in the surface 1–2 cm of many soils. They may serve as "early warning devices" for environmental stress. Tardigrades were found to be the most sensitive organism measured in a several-year study of the effects of dry deposition of SO_2 on litter and soil of a mixed-grass prairie ecosystem (Grodzinski and Yorks, 1981). They are thought to feed on algal cells and probably have a rather broad diet of various microbial-rich bits of soil organic matter. Tardigrades have also been observed to feed voraciously on nematodes when in culture (G. W. Yeates, personal communication, 1994).

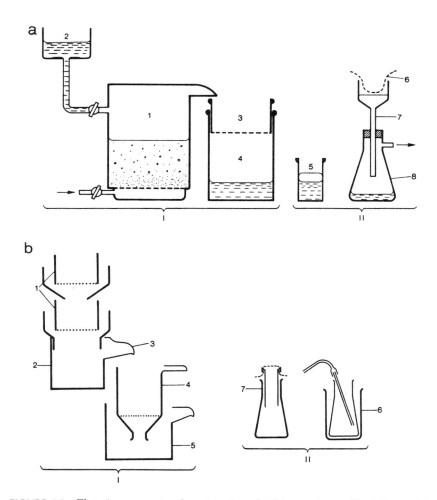

FIGURE 4.8   Flotation apparatus for extraction of soil invertebrates [Ladell's modified flotation apparatus). (a) Edwards and Dennis, 1962; (b) Salt (1953): I, separation of organic matter from the mineral components of the soil; II, separation of soil invertebrates and organic residue; (a): 1, flotation vessel with air supply from below; 2, container with salt solution; 3, sieve; 4, sedimentation tank; 5, glass oil separator; 6, gauze with the material under study; 7, glass funnel; 8, flask connected to suction pump; (b): 1, vessels with coarse and fine sieve; 2, container collecting the residue; 3, outlet; 4, final collecting sieve; 5, water tank; 6, water bath; 7, glass for oil–water flotation (from Kaczmarek, 1993).

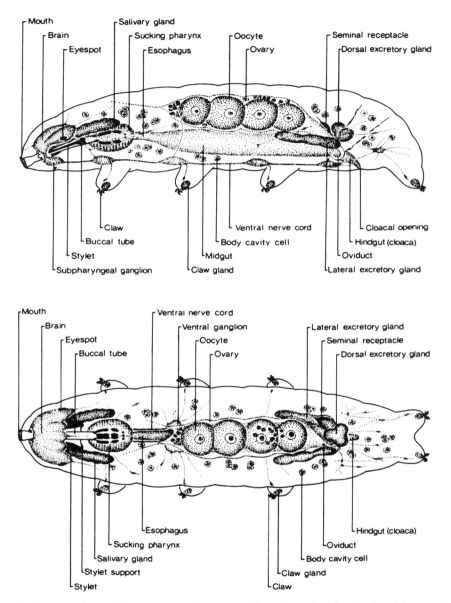

FIGURE 4.9 Eutardigrade internal anatomy (*Macrobiotus hufelandii*, female): lateral view (top) and dorsal view (bottom). (From "Soil Biology Guide," Dindal, D. L. Copyright © 1990 John Wiley & Sons, Inc. Reprinted by permission of John Wiley & Sons, Inc.).

Tardigrades have been found in large numbers (up to 2000 per 10 cm^2 of soil surface) and are particularly associated with lichens, mosses, liverworts, and rosette angiosperms (Nelson and Higgins, 1990). They also are found in very cold, dry habitats, such as the Antarctic dry valleys, where they feed on the particulate organic detritus brought in by windward movement of algal cells from lake ice at one end of the long valleys (D.W. Freckman, personal communication, 1994).

Tardigrades may be extracted from soils and various substrates by flotation and sieving through a 44 μm sieve or by the sucrose flotation and centrifugation technique used for extracting nematodes (Christie and Perry, 1951).

Collembola

The collembolans, mites and a variety of small insects are collectively known as microarthropods. These creatures are sampled in soil cores or litter bags (Chapter 5) and are extracted by heat gradient apparatus (Tullgren or Berlese funnels) or by flotation methods. With these techniques entire samples of soil are returned to the laboratory and the microarthropods are extracted by heat, desiccation, or mechanically (Edwards, 1991). Collections from these samples usually include representatives of collembolans, mites, small insects, and other small arthropods. Collembolans and mites are the most numerous microarthropods

Collembolans (Fig. 4.10) are primitive Apterygote insects. They have the common name of "springtails" from the fact that many of the species are able to jump by means of a lever (the furcula) attached to the bottom of the abdomen (Fig. 4.10). Collembolans are ubiquitous members of the soil fauna, often reaching abundances of 100,000 or more per m^2. They occur throughout the soil profile, where their major diet appears to be decaying vegetation and associated microbes. Models of soil food webs usually place collembolans as fungivores (e.g., Coleman, 1985; Hunt et al., 1987). However, like many of the soil fauna, collembolans in general defy such exact placement into trophic groups. Many collembolan species will eat nematodes when those are abundant. Some feed on live plants or their roots. One family (Onychiuridae) may feed in the rhizosphere and ingest mycorrhizae or even plant pathogenic fungi (Curl, 1979).

Eight families of collembolans occur in soils (see Table 4.2). In North America, identification beyond the family has been significantly improved by the publication of the Christiansen and Bellinger monograph (1980–1981), but many of our species remain unnamed, a challenge for microarthropod taxonomists.

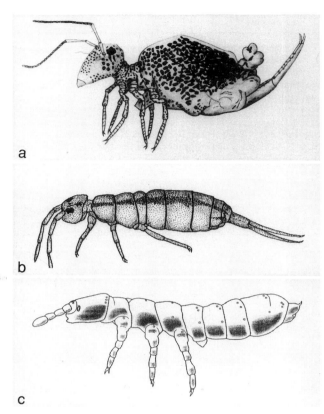

a

b

c

FIGURE 4.10   (a) A symphypleonid Collembolan (*Sminthurus burtcheri*) (Snider, 1969). (b) An arthropleonid Collembolan (*Isotomurus palustris*) (Snider, 1969). (c) An onychiurid Collembolan (Onychiuridae: *Protaphorura* sp.). Note the absence of the furcula (jumping apparatus) on the eyeless, soil-dwelling onychiurid, in contrast to the other litter-dwelling forms.

Many collembolans are opportunistic species, capable of rapid population growth under suitable conditions. Eggs are laid in groups. Collembolans become sexually mature with the fifth or sixth instar, but continue to molt throughout life. Some species have alternate stadia of growth and reproduction. Although many species are bisexual, some of the common species (e.g., *Folsomia candida*) are parthenogenic. Collembolan "blooms" are a phenomenon of late winter or early spring, when some species may appear in large numbers on the surface of snow banks ("snow fleas"), on the surface ice of pond water, or on lichen-covered granite outcrops. But they are, in essence, soil animals and are typical inhabitants of soil or soil-like habitats.

TABLE 4.2.   Key to Families of Collembola[a]

1a. Body elongate, with thoracic and abdominal segments distinct................................2
1b. Body globular, thoracic and first four abdominal segments fused...........................7
2a. Thoracic segments distinct; first segment with dorsal setae.....................................3
2b. Thoracic segments dissimilar. First thoracic segment usually hidden beneath
     second segment or, if visible, without dorsal setae...5
3a. Head hypognathus. First thoracic segment similar in size to second and third....
     ....Poduridae
3b. Head prognathus. First thoracic segment reduced compared to second and third...4
4a. Pseudocelli present. Eyes absent. Pigment usually lacking................Onychiuridae
4b. Pseudocelli absent. Eyes present or absent. Pigment usually present
     ....Hypogastruridae
5a. Abdominal segment IV usually longer than abdominal segment III........................
     ....Entomobryidae
5b. Abdominal segment IV subequal to or shorter than abdominal segment III..........6
6a. Scales present; dentes sometimes with multidentate spines............Entomobryidae
6b. Scales absent; dentes without such spines..Isotomidae
7a. Antennae shorter than head. Eyes absent...Neelidae
7b. Antennae longer than head. Eyes present or absent.........................Sminthuridae

---

Note. The study of Collembola involves slide mounts and a thorough knowledge of collembolan morphology. This simplified key may be useful for unmounted specimens of common species taken from soil samples. For more detailed keys, consult Christiansen and Bellinger (1980–1981) or Christiansen (1990).
[a]After Snider (1987).

Collembolans occur in aggregations. In samples of soils, collembolans are not found at random, but in groups. Aside from the statistical problems of assessment of population size, aggregations pose ecological questions. In laboratory investigations Christiansen (1970) and Barra and Christiansen (1975) analyzed collembolan responses to habitat variables (i.e., moisture and substrate) and food resources. Although these were important, the major variable seemed to be a behavioral one. It seems likely that collembolans possess aggregation pheromones (Christiansen *et al.*, 1992).

The importance of Collembola in soil systems defies easy assessment. Their large population sizes, together with rapid growth rates, suggests that they may have a significant impact on microbial dynamics, but this is not easy to demonstrate. Collembolans and their eggs may be important food items for a variety of other arthropods: predaceous mites, beetles, ants, and so forth.

Collembolan Feeding Habits

Models of food webs include collembolans as major food sources for a variety of predators. But their influence doubtless exceeds their role as generalized prey. Curl and Truelove (1986) argue persuasively that collembolans (among other fauna) are attracted to plant roots and are important in rhizosphere dynamics. In experiments, collembolans protected cotton plants from the root pathogen *Rhizoctonia solani* by selectively grazing that fungus from the plant roots. These rhizosphere inhabitants may prove to be effective biological control agents. Similarly, collembolans may be significant in the control of nematode populations (Gilmore, 1972). Feeding on nematodes does not appear to be selective; collembolans do not distinguish between saprophytic and plant parasitic nematodes.

Despite this tendency toward omnivory, current studies suggest that collembolans are primarily fungivorous. They may feed selectively on certain fungal species, thus influencing the fungal community and indirectly affecting rates of litter decomposition or nutrient cycling (Moore *et al.*, 1987). Selective grazing by the collembolan *Onychiurus latus* changed the outcome of competition between two basidiomycete decomposer fungi (Newell, 1984a,b), allowing an inferior competitor to prosper. Grazing upon fungi may actually increase general fungal activity in soils and stimulate fungal growth (Table 4.3). The relationship between fungal and collembolan population dynamics is not straightforward, however, since some collembolan species may actually reproduce more successfully on least favored foods (Walsh and Bolger, 1990). Collembola have been demonstrated to have complex interactions with several fungal species simultaneously. Cotton was grown in a greenhouse with four fungal species, the pathogen *Rhizoctonia solani* and three known biocontrol fungi (including two sporulating Hyphomycetes), and the rhizosphere inhabiting collembolan, *Proisotoma minuta*. The collembolan preferentially fed on the pathogenic fungus, and avoided the biocontrol fungi (Lartey *et al.*, 1994).

TABLE 4.3. Compensatory Growth of Fungi in Response to Collembolan Grazing[a]

Fungal species	Collembolan species	Growth relative to controls
Botrytis cinerea	*Folsomia fimentaria*	$-$[b]
Coriolus versicolor	*F. candida*	$-/+$[c]
Mortierella isabellina	*Onychiurus armatus*	$-$[b]
Verticillium bulbillosum	*O. armatus*	$+$
Penicillium Spinulosum	*O.armatus*	$+$
Field soil dilution	*Hypogastrura tullbergi*	$+$[b]
	F. regularis	$+$[b]

[a]After Lussenhop (1992).
[b]Bacteria were present.
[c]Increase with fungi grown on high nutrient medium, otherwise decrease.

Mites

The soil mites, Acari, chelicerate arthropods related to the spiders, are the most abundant microarthropods in many types of soils. In rich forest soil, a 100-g sample extracted on a Tullgren funnel may contain as many as 500 mites representing almost 100 genera. This much diversity includes participants in three or more trophic levels and varied strategies for feeding, reproduction and dispersal. Often, ecologists analyze samples by a preliminary sorting of mites into suborders. Identification of mites to the family level is a skill readily learned under the tutelage of an acarologist. Expert assistance is necessary for identifications of soil mites to genus or species. By combining slide mounts with examination of specimens in alcohol, reasonably accurate sorting of samples can be performed.

Four suborders of mites occur frequently in soils: the Oribatei, Prostigmata, Mesostigmata, and Astigmata. Slide mounts are required for accurate separations and reference to keys such as those in Krantz (1978). Mounting media suitable for mites, collembolans, and other microarthropods are described in Dindal (1990).

The soil mites are a subset of the Acari. Occasionally, mites from other habitats are extracted from soil samples. These include, for example, plant mites (red spider and false spider mites), predaceous mites normally found on green vegetation, and parasites of vertebrates or invertebrates. But the numerous species are the true soil mites. The oribatid mites (Oribatei) are the characteristic mites of the soil and are usually fungivorous or detritivorous. Mesostigmatid mites are nearly all predators on other small fauna, although some few species are fungivores and may become numerous. Acarid mites are associated with rich, decomposing nitrogen sources and are rare except in agricultural soils. The Prostigmata contains a broad diversity of mites with several feeding habits. Very little is known of the niches or ecological requirements of most soil mite species, but some tantalizing information is emerging (Petersen and Luxton, 1982; Walter et al., 1987).

Oribatei

Oribatid mites (Figs. 4.11–4.13) are an ancient group. They have the richest fossil record of any mite group, dating back to the Devonian, or 420–430 million years ago. Specimens from a Devonian site near Gilboa, New York contain organic matter visible in their guts, attesting to a long relationship between oribatid mites and decomposing vegetable matter (Norton et al., 1987). Some oribatid species are palearctic in distribution, being widely distributed in forest floors of Europe and North America.

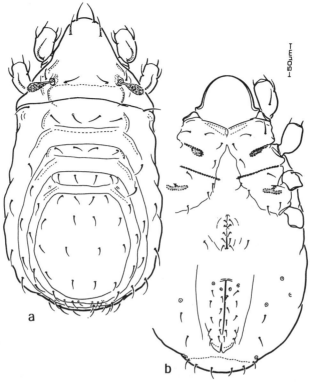

FIGURE 4.11 *Eueremaeus columbianus* (Berlese), tritonymph. (a) Dorsal aspect; (b) ventral aspect (from Behan-Pelletier, 1993).

A combination of three factors makes oribatid mites unique among the soil fauna. First, their sheer numbers are impressive. They are the most numerous of the microarthropods. Second, they possess juvenile polymorphism. Many immatures do not resemble adults. Unlike other mites, some immatures of the Oribatei are morphologically so dissimilar from the adult stadia that it is frequently impossible to correlate the two (compare Figs. 4.11 and 4.12). Despite the differences in morphology, they can usually be cultured on the same resources. Third, they reproduce relatively slowly in contrast to the other microarthropods. One or two generations per year are usual, and females do not lay many eggs. Oribatids are considered to be "K" specialists, and in contrast colllembolans would be "r" specialists or opportunistic species (MacArthur, 1972).

Oribatids differ from other microarthropods by having a calcareous exoskeleton. In this they resemble the millipedes, snails and isopods. The exoskeleton contains high calcium levels even in the primitive,

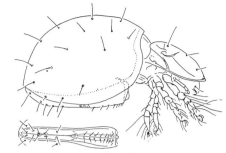

Genitoanal plates (ventral view)

FIGURE 4.12   A "box" mite of the oribatid family Phthiracaridae. For protection, the legs can be withdrawn beneath the hinged prodorsum (after Baker *et al.,* 1958).

lightly colored species (Fig. 4.13) (Todd *et al.*, 1974). The chemical form of the mineral deposits is principally calcium carbonate. Presumably, oribatids are able to sequester calcium by feeding on fungi. Senescent fungal hyphae contain crystals of calcium oxalate, which may be metabolized by the mites (Norton and Behan-Pelletier, 1991).

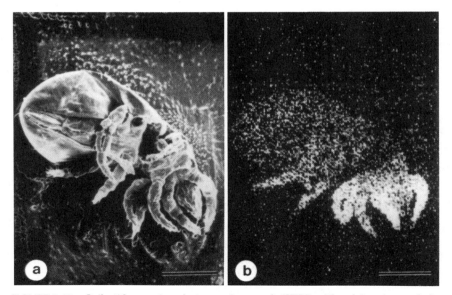

FIGURE 4.13   Oribatid scanning electron micrograph (SEM) with calcium in exoskeleton [a and b are micrographs of an oribatid mite (*Hypochthonius* sp.); a 100-μm marker is indicated at lower right.] (a) Image made by secondary electrons. (b) Map of calcium distribution made from X-ray image] (from Todd *et al.*, 1974).

Oribatid densities in forest soils are in the range of 250,000–500,000 per m² (Table 4.4). Coniferous forests typically support high numbers of oribatid mites, followed by deciduous hardwood forest, grassland, desert, and tundra. A general catalog of Oribatids of North America is given by Marshall *et al.* (1987). In arctic tundra and in some grassland or savannah habitats oribatids may be outnumbered by prostigmatic mites (see later). Cultivation of agricultural fields reduces oribatid populations to an average of about 25,000 per m². Population cycles in agroecosystems are often initiated by harvest and cultivation procedures, which change patterns of residue input into soils.

Populations of oribatid mites in forest floors show peaks of activity beginning in spring and continuing through the summer months, and again in mid-autumn (for those species producing two generations per year). Peaks of abundances of immatures show a gradual progression during the summer (Fig. 4.14) (Reeves, 1967). Population densities for many species are markedly higher during these months. Is there a sequence of oribatid species—a succession—in decomposing forest leaf litter corresponding to a succession of fungal species? Crossley and Hoglund (1962) found such a general relationship. In a detailed study, Anderson (1975) concluded that the dominant oribatid species rapidly colonized litterbags containing beech and chestnut leaf disks. The mites fed on the succession of fungi, but no succession of the mites themselves was demonstrated.

Feeding habits and nutrition of oribatid mites remain elusive, despite numerous detailed studies (e.g., Luxton, 1972, 1975, 1979). Oribatids in culture will eat a variety of substrates, but may feed differently under field conditions. In the simplest classification, oribatids are separated into feeders on fungi (microphytophages), on decomposing vegetable matter (macrophytophages), or on both

TABLE 4.4. Densities (Number per m²) of Soil Microarthropods in Four Forest Types[a]

Taxa	Mixed hardwood	Aspen woodland	Spruce forest	Scots pine forest
Oribatei	56,000	123,000	212,000	425,000
Prostigmata	[b]	25,000	96,000	250,000
Mesostigmata	1,500	7,400	14,000	8,600
Collembola	7,500	71,000	46,000	60,000

[a]Modified, with permission, from the *Annual Review of Entomology*, Volume 28, © 1983, by Annual Reviews Inc.

[b]Not estimated.

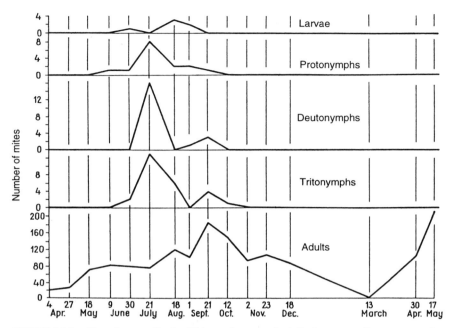

FIGURE 4.14   Abundances of oribatid immatures and adults by season [larvae, nymphs and adults of *Oppia subpectinata* (Oudemans)] (from Reeves, 1967).

(panphytophages) (Schuster, 1956). This classification has some utility, but more specific feeding habits can be identified. Some species, the phthiracarids or box mites, are largely macrophytophages. Many oribatids appear to be indiscriminate fungal feeders, ingesting fungal hyphae or fruiting bodies of a variety of species (Mitchell and Parkinson, 1976; Siepel and de Ruiter-Dijkman, 1993). Others may be selective. Adults of *Liacarus cidarus*, when offered a variety of foods, preferred the mold *Cladosporium* (Arlian and Woolley, 1970) (Table 4.5). Anderson (1975) studied competition between two generalist feeders, *Hermaniella granulata* and *Nothrus sylvestris*. When isolated, these two species used similar foodstuffs (based on gut analyses). But when kept together in soil litter microcosms, the two species changed their feeding and their utilization of habitat space. *Hermanniella* moved into the litter (A_o) layers while the *Nothrus* population increased in the F (A_i) layer.

Several lines of evidence—gut analyses, feeding trials in cultures, and chemical considerations—suggest that oribatid mites, as a group, are fungal feeders. Some exceptions exist, such as the phthiracarid group, which may bore into coniferous needles or twigs and ingest the

TABLE 4.5.   Feeding Behavior of Adult *Liacarus cidarus* When Offered Different Resources in Laboratory Cultures[a]

Resource	Heavy feeding	Light feeding	Intermittent feeding	Failed to feed
Lichens	—	X	—	—
Yeast and sugar	—	—	X	—
Pine cone scales	—	X	—	—
Cladiosporium	X	—	—	—
Aspergillus	—	—	—	X
"Mushrooms"	—	—	X	—
Potato dextrose agar and *Cladiosporium*	X	—	—	—
Potato dextrose agar	—	X	—	—
Pine litter	—	—	X	—
Yeast	—	—	—	X
Trichoderma	—	—	—	X

[a]After Arlian and Woolley (1970).

spongy mesophyll. These species may contain a gut flora of bacteria which allow them to digest decomposing woody substrates. Possibly, their nutrition is derived from bacteria or fungi embedded in the woody tissue. Examination of gut contents for most other oribatids shows that they feed primarily on fungi. A survey of 25 species from the North American arctic found that over 50% were panphytophagous (feeding on microbes and plant debris), but that nearly all contained fungal hyphae or spores. Similarly, a study of Irish species found that 15 of 16 species were generalist feeders, having both fungi and plant remains in their guts (Behan and Hill, 1978; Behan-Pelletier and Hill, 1983). Occasional fragments of collembolans were discovered in some of the guts. Morphology of the chelicerae seems related to feeding type (Kanoko, 1988). Xylophagous (wood-feeding) species have large, robust chelicerae. Fragment feeders are generally smaller and have smaller chelae.

Further refinement of guild designations for oribatid mites has been made based on their digestive capabilities as evidenced by their cellulase, chitinase, and trehalase activity in field populations (Siepel and de Ruiter-Dijkman, 1993). Using these three enzymes, it was possible to recognize five major feeding guilds: herbivorous grazers, fungivorous grazers, herbo–fungivorous grazers, fungivorous browsers, and opportunistic herbo-fungivores. In this classification, grazers are species that can digest both cell walls and cell contents; browsers can digest only cell contents. Siepel and de Ruiter-Dijkman also recognized two minor guilds: herbivorous browsers and omnivores. Further work with other groups of gut enzymes may provide additional information concerning resource utilization, and fungal feeding in particular, by guilds of oribatid mites.

Prostigmata

The Prostigmata contains a large array of soil species (Fig. 4.15). Many of these species are predators, but some families contain fungal-feeding mites and these may become numerous. Like the oribatids,

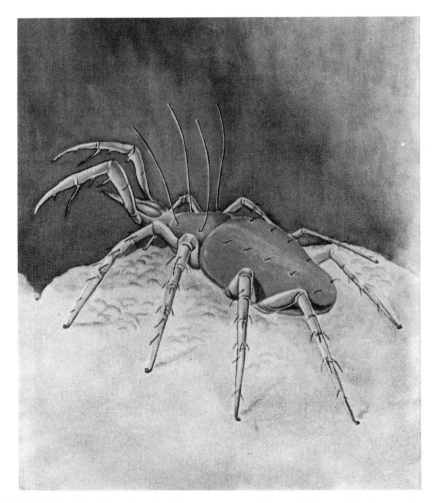

FIGURE 4.15   Prostigmatid mite [*Cunaxa taurus* (Kramer) Thomas M. Evans] (from Baker and Wharton, 1952).

prostigmatic mites are an ancient group with fossil representatives from the Devonian era. The fungal-feeding species are opportunistic and are able to reproduce rapidly following a disturbance or a sudden shift in resources. Small species of Prostigmata are the common mites of Antarctic soil surfaces, of drained lake beds with algal blooms, of plowed and fertilized agricultural fields, of southwest deserts after rains, of tidal marshlands or burned prairie soils, and so forth (Lussenhop, 1976; Luxton, 1967, 1981b; Perdue and Crossley, 1989; Seastedt, 1984a; Strandtmann, 1967; Tevis and Newell, 1962). Species in the families Eupodidae, Tarsonemidae, Nanorchestidae, and some of their relatives feed on algae or fungi; their populations may grow rapidly to large sizes. In these situations the Prostigmata may become more numerous than the oribatid mites. In general they are more numerous in temperate than in tropical or subtropical habitats (Luxton, 1981b).

Many types of fungal-feeding Prostigmata have small, stylet chelicerae and may simply pierce fungal hyphae. Some of the smaller predatory species may utilize fungi on occasion. In general, the larger predaceous Prostigmata feed on other arthropods or their eggs; the smaller species are nematophagous. Some Prostigmata have well-defined patterns of predation. The "grasshopper mite," *Allothrombium trigonum*, feeds exclusively on grasshopper eggs; the larval stages of the mite are parasitic on grasshoppers. The large red "velvet mites" (*Dolicothrombium* species), which erupt in numbers following desert rains, are predaceous on termites. The pestiferous "chiggers" are the larval stages of mites in the family Trombiculidae; the adults are predaceous on collembolans and their eggs. The smaller predaceous Prostigmata appear to specialize on nematodes (Walter, 1988), although some occasionally feed on fungi. Some of these mites may ingest particulate matter whereas others may ingest fluids only.

Mesostigmata

The Mesostigmata (Fig. 4.16) contains fewer species than oribatid or prostigmatic mites. Soil species are almost all predators. A few species (in the Uropodidae, for example) are fungal feeders and become somewhat numerous in agroecosystems (Mueller *et al.*, 1990). Mesostigmatic mites are not as numerous as oribatids or prostigmatic mites, but are universally present in soils and may be important predators. As with the Prostigmata, the larger species tend to feed on small arthropods or their eggs; the smaller species are mainly nematophagous. Walter and Ikonen (1989) found that mesostigmatic mites were the most important predators of nematodes in grasslands of the western United States. In contrast, they found that all but the smallest Prostigmata were predators on arthropods or their eggs. Of

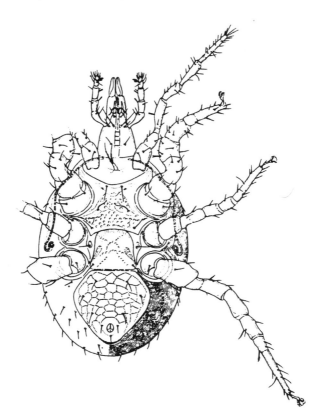

FIGURE 4.16   A mesostigmatid mite, family Macrochelidae. These mites are abundant in composted manure or cattle dung, where they feed upon nematodes and eggs of flies. Reprinted from Baker *et al.*, 1958, with permission from the Acarology Laboratory.

63 species of mesostigmatic mites tested, only 6 did not readily feed on nematode prey. The mites each consumed three to eight nematodes per day. Western grassland soils have little surface plant litter. Forest floors, with abundant surface litter, contain a larger spectrum of mesostigmatic mites. The forest litter inhabitants (families Veigaiidae and Macrochelidae, for example) are bigger species predaceous on arthropods or their eggs. Mesostigmatic mites of the mineral soil layers are the smaller, colorless Rhodacaridae and relatives, and are nematophagous. Many species of Mesostigmata have a close relationship with various insect species, a relationship which often includes the soil environment (Hunter and Rosario, 1988). Several genera in the Cohort Gamasina are also considered useful as bioindicators of habitat and soil conditions (Karg, 1982).

Astigmata

The Astigmata (Fig. 4.17) are the least common of the soil mites, although they may become abundant in some habitats (Luxton, 1981a). The free-living Astigmata are favored by moist environments high in organic matter. The Astigmata contain some important pests of stored grain. They become abundant in some agroecosystems following harvest or after application of rich manures. In Georgia piedmont agroecosystems, Perdue (1987) found a marked increase in Astigmata following autumnal harvest and tillage (Fig. 4.18). Incorporation of residues and winter rains produced moist, organic residues suitable for the mites. The springtime plowing, under drier conditions, did not lead to increases of astigmatic mites. Tomlin (1977) described a buildup of astigmatic mites following pipeline construction in Ontario, Canada. The mites were associated with accumulations of residue under moist conditions.

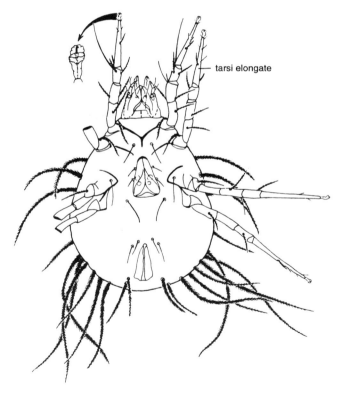

FIGURE 4.17   Astigmatid mite [*Glycyphagus domesticus* (DeGeer) (Oregon), venter of female with detail of pretarsus I] (from Krantz, 1978).

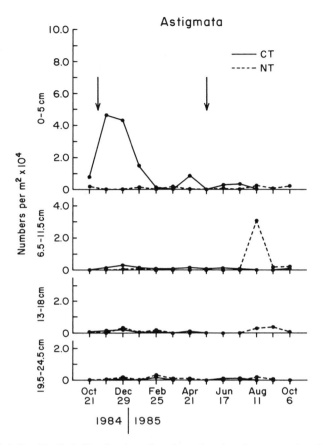

FIGURE 4.18   Vertical distribution of astigmatic mites in conventional and no-tillage agroecosystems. Arrows indicate autumn and spring dates for mowing, tillage, and planting. Numbers increased under conventional tillage following autumn tillage, but not following spring tillage (from Perdue and Crossley, 1989).

We have found astigmatic mites as contaminants in some Tullgren extractions. When fresh agricultural products are stored in the laboratory, large populations of Astigmata may develop and may wander into Tullgren funnels during extractions. Similar population excursions of prostigmatic mites (family Cheyletidae) have also yielded extensive contamination of Tullgren samples. It is good practice to operate some empty "control" funnels to check for the possibility of wandering microarthropods in the funnel room.

Other Microarthropods

In addition to mites and collembolans, Tullgren extractions contain a diverse group of other small arthropods. Although not numerous in comparison to mites and collembolans, they may have abundances of several thousand per square meter. Collectively, the "other" microarthropods have relatively small biomasses and probably have no significant impact on soil ecology. Such a judgment may be premature in view of the general lack of information about their ecologies.

Protura

Protura are small, wingless, primitive insects which lack antennae and eyes. They occur in a variety of soils worldwide, often associated with plant roots and litter. Their feeding habits remain unknown. Observations that they feed on mycorrhizae have not been verified.

Diplurans

Diplurans are small, elongate, delicate, primitive insects. They have long antennae and two abdominal cerci. Most diplurans are euedaphic, but some are nocturnal cryptozoans, hiding under stones or under bark during the day. They occur in tropical and temperate soils in low densities. In Georgia piedmont agroecosystems we have sampled dipluran populations with Tullgren extractions, finding populations of approximately 500 per m^2.

Two families are found in soils, separable by their abdominal cerci. Campodeidae have filiform cerci; whereas Japygidae have cerci modified as pinchers. The japygids are predators on small arthropods (such as collembolans), nematodes and enchytraeids. The cerci are used in capturing prey. Campodeids are saprophagous, feeding on dead or decaying organic matter. These animals are adapted for life in the soil by their elongate narrow form, sensory antennae, and sensory cerci.

Pseudoscorpions

Pseudoscorpions are minute copies of their more familiar relatives, the scorpions, except that they lack tails and stingers. Pseudoscorpions are small cryptozoans, hiding under rocks and bark of trees, but they are occasionally extracted in Tullgren collections. Pseudoscorpions are more numerous in tropical and subtropical regions. They are predaceous on small arthropods, nematodes and enchytraeids.

Symphylids

Symphylids are small, white, eyeless, elongate, many-legged invertebrates which resemble tiny centipedes. They are part of the true eudaphic fauna, occurring in forest, grassland and cultivated soils. Symphylids are omnivorous and can feed on the soft tissues of plants or animals. Some species reach pest status in greenhouse soils where they feed on roots of seedlings (Edwards, 1990). Symphylids have silk glands near the end of the abdomen. The function of silk strands for these soil dwellers is obscure.

Pauropoda

Pauropoda are tiny (1.0–1.5 mm long) terrestrial myriapods with 8–11 pairs of legs and a distinctive morphological feature: branched antennae. They are white to colorless, are blind, and are members of the true euedaphic fauna. Pauropods occur in soils worldwide but are not well known. They are commonly collected in Tullgren extractions but are seldom numerous, usually fewer than 100 per m². In forests they inhabit the lower litter layers, F layers, and mineral soil; they also occur in agricultural soils. It is generally assumed that pauropods are fungus feeders, but they may also be predaceous. Little information has been accumulated about their biology or ecology (Scheller, 1990).

Small spiders and centipedes, occasional small millipedes, insect larvae and adult insects occur in soil cores extracted on Tullgren funnels. Most of these are better sampled as macroarthropods using hand sorting or trapping methods. Some insects (small larvae of carabid and elaterid beetles, thrips, pselaphid beetles, tiny wasps) are sometimes numerous enough to be effectively sampled from soil cores. Ants and termites, social insects, require special sampling considerations.

Enchytraeidae

These organisms are a little-studied (in North America) family of Oligochaeta, also known as "potworms." The body size is usually 10–20 mm long, and they are a fairly simple group anatomically. A total of some 600 species are now known worldwide (Dash, 1990). Enchytraeids are anatomically similar to the earthworms except for the miniaturization of features overall. They possess setae (with the exception of one genus) and a clitellum in segments XII and XIII, which contains both male and female pores. These creatures are hermaphroditic, similar to earthworms (Fig. 4.19). Numerous investigators have noted that finely divided plant materials, often enriched with fungal hyphae, are a principal portion of the diet of Enchytraeids. Reynoldson (1939) found that Enchytraeids of sewage

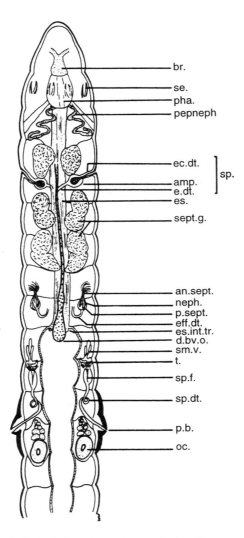

FIGURE 4.19 Morphological characters of an enchytraeid worm. amp., ampulla: an. sept., ante-septal; br., brain; d.bv.o., dorsal blood vessel origin; ec.g., ectal gland; eff.dt., efferent duct; e.op., ental opening; es., esophagus; es.int.tr., esophageal intestinal transition; m.pha., muscular pharynx; neph., nephridia; oc., oocyte; pha., pharynx; p.b., penial bulb; pepneph., peptonephridia; p.sept., postseptal; se., setae; sept.g., septal gland; sm.v., seminal vesicle; sp., spermatheca; sp.dt., sperm duct; sp.f., sperm funnel; t., testes (From the "Soil Biology Guide" Dindal, D. L. Copyright ©1990 John Wiley & Sons, Inc. Reprinted by permission of John Wiley & Sons, Inc.).

beds ingested algae, fungi, and bacteria. Didden (1990) suggests that Enchytraeids feed predominantly on fungi, at least in arable soils. As with several other members of the soil mesofauna, the mixed micro-biota which occur on decaying organic matter, either litter or roots, are probably an important part of the diet of these creatures. The remaining portions of the soil organic matter, after the processes of ingestion, digestion and assimilation, enter the slow turnover pool of soil organic matter. Zachariae (1963, 1964) studied the nature of enchytraeid feces and found that they had no identifiable cellulose residues. In addition, Zachariae suggested that so-called "collembolan soils," said to be dominated by collembolan feces (particularly low pH mor soils), were really formed by Enchytraeidae. Studies have found mycorrhizal hyphae in the fecal pellets of Enchytraeids from pine lit-ter (Fig. 4.20) (Ponge, 1991). Zachariae (1964) suggested that Enchytraeids were consuming and further processing collembolan fecal pellets. There is also the strong likelihood that enchytraeids in field sites are preferentially ingesting earthworm castings (J.M. Springett, personal communication, 1994.).

Enchytraeid numbers range from 10,000 to 64,000 per m^2 in *Pinus radiata* plantations in New Zealand (Yeates, 1988) and from 4000 to 14,000 per m^2 in agricultural plots in the Georgia piedmont (Coleman *et al.*, 1994a). Total numbers of enchytraeids are somewhat higher (ca. 20,000–30,000 per m^2) in surface layers of deciduous forest soils in the southern Appalachians of North Carolina (van Vliet *et al.*, 1995).

MACROFAUNA

Macroarthropods

These species are not as effectively sampled with Tullgren extrac-tions as are microarthropods, but require other methods. Large soil cores (10 cm in diameter or greater) may be appropriate for euedaphic species; arthropods are recovered from them using flotation tech-niques (Edwards, 1991) or hand sorting. Litter- and surface-dwelling arthropods are often sampled with pitfall traps (Gist and Crossley, 1973). Mechanical or hand sorting of litter is more time consuming but yields more accurate estimates of population size. In rare instances, capture-mark-recapture methods have been used to esti-mate population sizes, but the assumptions for this procedure are violated more often than not with soil macroarthropods (Southwood, 1978).

Many of the macroarthropods are members of the group termed "cryptozoa," a group consisting of animals which dwell beneath stones,

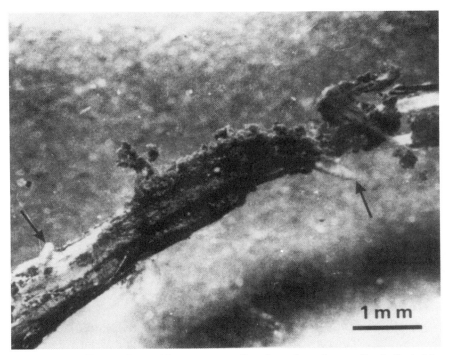

FIGURE 4.20   Two enchytraeid worms tunneling through a pine needle, indicated by arrows (fecal pellets have been deposited on the outside). F_1 layer (modified from Ponge, 1991).

logs, under bark, and in similar protected or hidden habitats (Cole, 1946). Cryptozoans typically emerge at night to forage. The cryptozoan fauna is poorly defined and is not an ecological community in the usual sense of the term. The concept remains useful, however, for identifying a group of invertebrates with similar patterns of habitat utilization.

Isopods

Terrestrial isopods (Fig. 4.21) are typical cryptozoa, occurring under rocks and in similar habitats. Although they are distributed in a variety of habitats, including deserts, they are susceptible to desiccation. Adaptations to resist desiccation include nocturnal habits, the ability to roll up into a ball, low basal respiration rates and restriction of respiratory surfaces to specialized areas. Considered to be general saprovores, isopods can feed on roots or foliage of seedlings. Isopods possess heavy, sclerotized mandibles and are capable of considerable

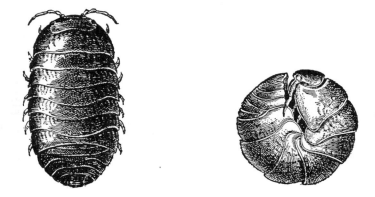

FIGURE 4.21   A terrestrial isopod, *Armadillidium vulgare*. Left, extended; right, rolled into a ball (Metcalf and Flint, 1939).

fragmentation of decaying vegetable matter. They display some selectivity in preferences for different leaf species.

Diplopoda

Millipedes (Diplopoda, Fig. 4.22) are a group of widely distributed saprophages. They are major consumers of organic debris in temperate and tropical hardwood forests where they feed on dead vegetable matter. Millipedes are also inhabitants of arid and semiarid regions, despite their dependence on moisture. Millipedes lack a waxy layer on their epicuticle and are subject to rapid desiccation in environments with low relative humidity. Some are true soil forms whereas others seem restricted to leaf litter or to cryptozoan habitats. They can be loosely grouped into (1) tubular, round-backed forms such as the familiar *Narceus*; (2) flat-backed forms (many Polydesmid millipedes); and (3) pillbug types, which roll into a ball. Millipedes range widely in length. Typical North American forms are 5–6 cm in length whereas tropical ones may reach nearly 20 cm in length.

Millipedes become abundant in calcium-rich, high rainfall areas in tropical and temperate zones. The southern Appalachian mountains of the eastern United States support a large millipede population. Millipedes can be important in calcium cycling. They have a calcareous exoskeleton and, because of their high densities, they can be a significant sink for calcium. Millipedes are major consumers of fallen leaf litter and may process some 15–25% of calcium input into hardwood forest floors. In desert areas, millipedes are active following rains, especially in desert shrub communities. They avoid hot, dry conditions by concealment under vegetation or debris (Crawford, 1981).

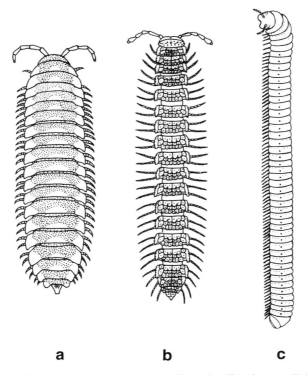

a   b   c

FIGURE 4.22 Representatives of three families of millipedes: (a) Xylodesmidae (b) Polydesmidae (c) Spirobolidae (Kevan and Scudder, 1989).

Millipedes appear to be selective feeders, avoiding leaf litter high in polyphenols and favoring litter with a high calcium content (Neuhauser and Hartenstein, 1978). Some millipedes are obligate coprophages. When McBrayer (1973) cultured millipedes in containers which excluded their feces, the millipedes lost weight. When a small tray containing feces was added to the cultures, the millipedes consumed it and prospered. Such obligate coprophagy indicates a close relationship with bacteria necessary for digestion of vegetable material. It is not known whether millipedes possess a unique gut flora of microbes.

Chilopoda

Centipedes (Chilopoda) are common predators in soil, litter, and cryptozoan habitats (Fig. 4.23). They are all elongate, flattened, active forms. Centipedes occur in biomes ranging from forest to desert. The

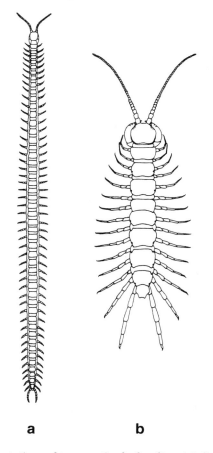

a   b

FIGURE 4.23   Representatives of two centipede families: (a) Geophilidae (*Geophilus proximus* Koch) (b) Lithobiidae [*Lithobius forficatus* (L.)].

large desert centipedes (Scolopendromorpha) are some 15 cm long; tropical centipedes may exceed 30 cm. Lithobiids are the common brown, flat centipedes of litter in hardwood forests. The elongate, slim geophilomorph centipedes are euedaphic in forest habitats. Like the millipedes, centipedes lose water through their cuticles at low relative humidities. They avoid desiccation by seeking moist habitats and by adjusting their diurnal activities to humid periods.

All centipedes are predators, but may ingest some leaf litter on occasion—it can sometimes be seen in their guts. Centipedes are fast runners and actively pursue and capture small prey such as collembolans.

Scorpions

The scorpion, the archetypic generalized arachnid with its long, segmented sting-bearing abdomen and chelate palpi, needs no description. It was obvious to the ancients; it is the only zodiacal sign bearing the name of a soil organism. Scorpions (Fig. 4.24) are inhabitants of warm, dry, tropical and temperate regions but reach their greatest diversity in deserts. They are highly mobile predators of other arthropods, lizards, mice, and birds. They are also cannibalistic to an unusual extent (Williams, 1987). Typical cryptozoans, scorpions hide under rocks, logs, or in crevices during the day and emerge at night to feed. The impact of scorpions on their ecosystems is unknown. They are not numerous, but in desert ecosystems they may be dominant predators (Polis, 1991).

Araneae

Spiders (Araneae, Fig. 4.25) are another familiar group of carnivores. They are found in all terrestrial environments except truly polar (Arctic/Antarctic) regions. Many species are aboveground predators, but some are cryptozoans and some forms are euedaphic. Some of the small litter inhabiting spiders could be considered microarthropods. Spiders may be active hunters or "sit and wait" dwellers in retreats. Wolf spiders (Lycosidae) are wandering predators in leaf litter and on soil surfaces, and are often captured in pitfall traps. They are conspicuous ground-dwelling predators in agroecosystems.

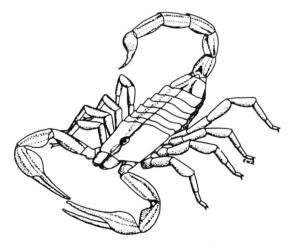

FIGURE 4.24   A scorpion, shown 1½ times its actual size (from Borror *et al.*, 1981).

FIGURE 4.25   A typical wolf spider, *Lycosa communis* (Lycosidae) (from Wise, 1993).

Little is known about the ecology of soil- and litter-dwelling spiders. Most information about spiders comes from studies of web-spinning species or the jumping spiders (Attidae) in vegetation. Large numbers of spiders are found in deciduous forest leaf litter (about 100 per m²). Their habitat usage and prey selection are not well known. Spiders in laboratory microcosms show some prey selectivity, but leaf litter spiders doubtless feed opportunistically.

The impact of spiders on their ecosystems is not well known. Their effectiveness as biological control agents has been discounted because of their slow reproduction. Spiders are strongly territorial, with complicated mating rituals, adaptations which tend to hold down population sizes even when prey are abundant (Wise, 1993).

Opiliones

Harvestmen (Opiliones) are delicate, shy forms which are among the largest arachnids in woodlands. They have no venom glands, yet are considered to be largely predaceous. Some species occur high in foliage, others in subcanopy, some on soil surface, and some (smaller forms) in litter layers. Opilionids are slow reproducers, usually with one generation per year. They are active predators in the daylight but seem to be primarily crepuscular (active dawn and dusk).

Other Apterygota

Soil systems are the habitat for primitive, wingless insects: the Apterygota. These include the collembolans, diplurans, and proturans, which were already discussed among the microarthropods. An additional order, the Thysanura, is numbered among the soil fauna. Their size, approximately 5 mm long and larger, places them among the macroarthropods.

Thysanurans are represented in leaf litter and soils by the family Machilidae. Easily recognized by their hump-backed appearance and their three long abdominal cerci, machilids are true cryptozoans. They are adapted for life on the soil surface: they are cryptically colored, have the ability to leap, and can tolerate moderate to low moisture conditions. Their principal habitat appears to be rocky soil surface. Authorities list machilids as saprovores. They are presumed to feed on algae, lichens, decaying vegetable debris, and associated fungus (Ferguson, 1990).

In some habitats, machilids are relatively numerous. On the bare granite outcrops of the Georgia piedmont, we found 50–100 individuals per m^2 at night, making them among the more numerous arthropods in these relatively depauperate ecosystems. They hide in cracks in the rocks during the day and forage on the lichen-covered rock surfaces at night.

Pterygota. Many of the higher insect orders, the Pterygota, are participants in soil systems. Some, e.g., grasshoppers, contribute eggs as inputs to soil food webs, e.g., as sustenance for trombidiid mites. Canopy caterpillars descending via silken threads to the soil to pupate may successfully avoid tachinid flies only to fall into the waiting clutches of lycosid spiders or carabid beetles. Several minor orders of insects contain true soil inhabitants, e.g., the Dermaptera (earwigs), Psocoptera, Zoraptera, and Thysanoptera.

Three orders of higher insects contain species which have major impacts on soil systems, their structure, chemistry, and food webs. These are the Isoptera (termites), Hymenoptera (ants and wasps), and Coleoptera (beetles). We cannot provide an adequate treatment of their biology and ecology in this text; each order would require its own volume. A brief overview is given here.

Isoptera. The Isoptera (Fig. 4.26), the termites, are among the most important of soil fauna in terms of their impact on soil structure and on decomposition processes. Termites are social insects with a well-developed caste system. Through their ability to digest wood, they have become economic pests of major importance in some regions of the world (Lee and Wood, 1971).

Termites in the primitive families, such as Kalotermitidae, possess a gut flora of protozoans, which enables them to digest cellulose. Their normal food is wood that has come into contact with soil. Most species of termites construct runways of soil, and some are builders of spectacular mounds (Fig. 4.27). Members of the phylogenetically

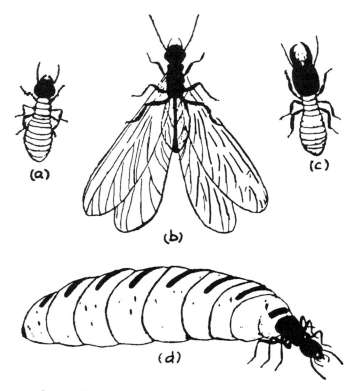

FIGURE 4.26   Castes of termites. (a) Worker (b) winged reproductive (c) soldier and (d) queen (Veeresch and Rajagopal, 1983).

advanced family Termitidae do not have protozoan symbionts, but possess a formidable array of microbial symbionts (bacteria and fungi) which enable them to process and digest the humified organic matter in tropical soils and to grow and thrive on such a diet (Breznak, 1984; Bignell, 1984).

A number of inquilines (organisms existing in and sharing common space) occur in termite nests—ants, collembolans, mites, centipedes, and beetles which have become morphologically specialized for that habitat.

Although termites are mainly tropical in distribution, they occur in temperate zones and deserts as well. Termites have been called the tropical analogs of earthworms since they reach large abundances in the tropics and process large amounts of litter. Three nutritional categories include wood-feeding species, plant- and humus-feeding species, and fungus growers. This latter group lacks intestinal symbionts and depends on cultured fungus for nutrition.

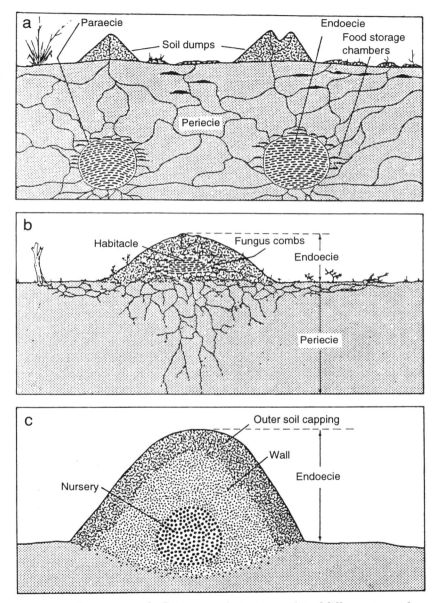

FIGURE 4.27   Termite mounds: Diagrammatic representation of different types of concentrated nest systems. (a) *Hodotermes mossambicu* (b) *Macrotermes subhyalinus*, and (c) *Nasutitermes exitiosus* (from Lee and Wood, 1971).

Termites are one of the three major earth-moving groups of invertebrates (the other two are earthworms and ants). Mound-building termite species have a major impact on the distribution and composition of soil mineral and organic matter. Annual crops may grow poorly in soil from termite mounds as it is usually poorer in nutrients and more compact than the surrounding soil.

Termites were thought to contribute a significant proportion of the total biogenic methane in the biosphere (Zimmerman *et al.*, 1982), but it is now thought to be relatively minor (Schneider, 1989).

Hymenoptera. The order Hymenoptera is one of the largest orders of insects. It contains two groups of soil insects of large importance: ants (Fig. 4.28) and ground-dwelling wasps (Fig. 4.29). Ants are among the most successful insects, occurring in nearly every terrestrial habitat (Hölldobler and Wilson, 1990). All ants are colonial, social insects with at least three castes: queens, males, and workers. The majority nest in soil. The feeding habits of ants are highly variable and include carnivores, saprovores, seed predators, plant secretions,

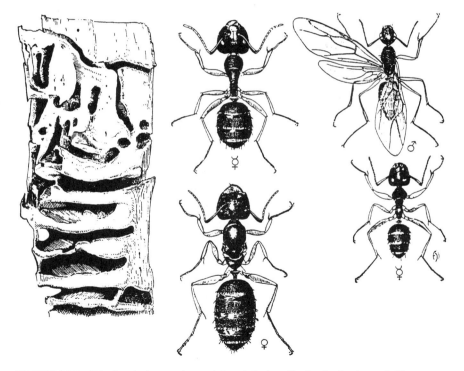

FIGURE 4.28   Wood ants (carpenter ants) and their galleries in dead wood. Shown are a large and a small neuter worker, a winged male and a wingless egg-laying female (Henderson, 1952).

FIGURE 4.29   A digger-wasp (family Sphecidae). These solitary wasps usually prepare nests in the soil, which they provision with arthropod prey before depositing eggs (Pratt and Stojanovich, 1967).

aphid secretions, and so forth. Many are opportunistic omnivores. Ants influence the structure of soil systems by excavating nests and depositing lower soil layers on the surface, a phenomenon easily observed in mound-building ants. Nutrients are redistributed in the process of nest maintenance. Ants are also major predators on microarthropods. Because of their social nature, ant population sizes are difficult to measure.

Most of the solitary wasps in the superfamily Vespoidea construct nests in the soil. The adult female wasp first constructs a small nest cavity. Then a suitable prey item (another insect or a spider) is located, which the wasp then stings to paralyze it and hauls it to the nest. An egg is laid on the paralyzed victim, and it is then entombed. Some of the social wasps, especially *Vespula* spp., nest in the ground. Natural cavities, such as abandoned rodent burrows, are often used as nesting sites. Vespids are carnivorous, feeding their larvae on captured prey (insects or spiders), although adult wasps generally feed on nectar, sap, or similar juices (Michener and Michener, 1951).

Coleoptera. The Coleoptera is the largest order of insects. Beetles exhibit wide variation in form, size, function, and distribution. They are worldwide in distribution and are found in every habitat except oceans. Thomas Henry Huxley remarked that the Almighty must have loved beetles because he made so many kinds of them.

Beetles in the soil can be separated into predatory, phytophagous, and saprophagous forms. They include permanent residents, temporary residents and transients. Some of the more abundant families include:

Carabidae (Fig. 4.30), the ground beetles. Predators and seed eaters.

Tenebrionidae (Fig. 4.31), similar to carabids, typical beetles of deserts.

Staphylinidae (Fig. 4.32), rove beetles. Predators.

Scarabaeidae (Fig. 4.33), scavengers, dung beetles and root-feeding herbivores.

Cicindelidae (Fig. 4.34), tiger beetles. Adults are active, flying predators on exposed soil surfaces. Predaceous larvae live in retreats in the soil.

Elateridae (Fig. 4.35), larvae ("wireworms") are important root feeders in agroecosystems and forests.

Beetles bridge the gap between mesofauna and macrofauna. They are probably of more importance for their phytophagous activities above ground than for their participation in soil food webs. The predatory activities of ground surface beetles are significant in agricultural systems since they prey on pest species. Beetles are important agents in the reduction of dung and animal carcasses, and in the early decomposition of wood on the forest floor (Wallwork, 1982; Hanula, 1995).

Oligochaeta: Earthworms

Earthworms are the most familiar members of the soil fauna and are among the most significant ones. Over a century ago Darwin's last book, "The Formation of Vegetable Mould through the Actions of Worms, with Observations of Their Habits," called attention to the beneficial effects of earthworms: "It may be doubted whether there are many other animals which have played so important a part in the history of the world, as have these lowly organized creatures" (Darwin, 1881). Today, numerous studies have established the importance of earthworms as biological agents in soil formation, organic litter decomposition, and redistribution of organic matter in the soil (Edwards and Lofty, 1977; Lee, 1985; Hendrix, 1995).

Earthworms are long-lived animals but field data on longevity are rare. Life spans of common species have been estimated to be between 1 and 10 years. Earthworms are hermaphroditic. Their ova are produced in cocoons, usually during the warmer months, and hatchlings reach maturity in about a year (Figs. 4.36a, 4.36b). Populations sampled from field sites reveal a tapered pyramid of numbers, with a base of young forms and few older adults at the top.

Earthworms are members of the class Oligochaeta, which also includes the Enchytraeids. Some 20 families of earthworms are usually recognized (Reynolds and Cooke, 1993), including the common Lumbricidae and the equally important Megascolecidae. The latter family includes some widespread tropical and subtropical species. The familiar Lumbricidae contains the most important species in the northern hemisphere. Lumbricid species are "camp followers," which have become widely distributed by agricultural practices and by other human activities which displace large amounts of soil. In North America native earthworm species have been largely displaced by European species, except in refugia of native vegetation (Fender, 1995). The taxonomy of North American earthworms is poorly developed, and many species continue to be recognized and named (Fragoso *et al.*, 1995; James, 1995). Their biologies also are poorly known.

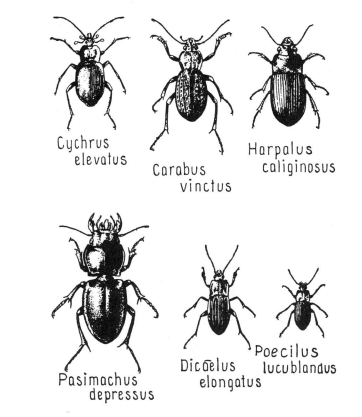

FIGURE 4.30 Illustrations of ground beetle species in the family Carabidae (Lutz, 1948).

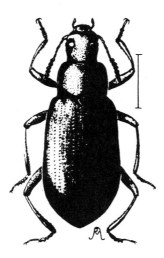

FIGURE 4.31   Tenebrionidae, dorsal view of *Alobates* sp. (false mealworm), 20–23 mm. Adults and larvae are found throughout the year beneath bark and logs, where they feed on other insects (from Arnett, 1993).

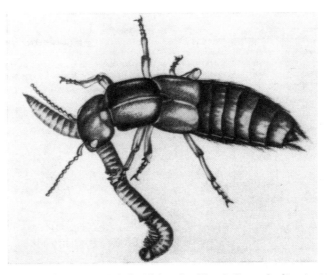

FIGURE 4.32   A predaceous staphylinid beetle (*Staphylinus badipes)* attacking a millipede (*Ophyiulus pilosus*) (Snider, 1984).

Canthon laevis

Geotrupes blackburnii

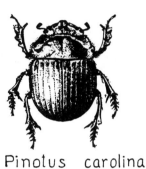

Pinotus carolina

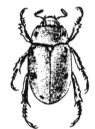

Trox suberosus

Pelidnota punctata

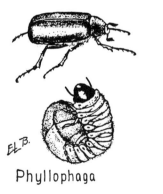

Phyllophaga

Macrodactylus subspinosus

FIGURE 4.33 Some scarabaeid beetles: *Canthos laevis*, a tumble-bug with a ball of dung in which an egg is laid; *Geotrupes blackburnii* and *Pinotus carolina*, also dung beetles. *Trox suberosus* lays eggs in carrion. Larvae of *Pelidnota punctata* live in decaying oak or hickory stumps. The J-shaped larvae of *Phyllophaga* species feed upon roots of plants (Lutz, 1948).

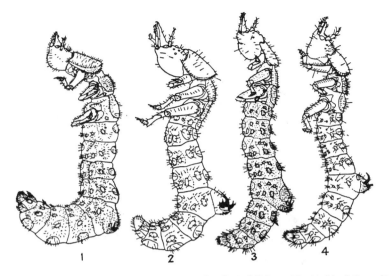

FIGURE 4.34   Larvae of tiger beetles, family Cicindelidae. (1) *Amblychila cylindriformis*; (2) *Omus californicus*; (3) *Tetracha carolina*; (4) *Cicindela limbalis*. The predaceous larvae lie in wait in vertical burrows with their heads flush with the soil surface, and held in place by hooks on the hump protruding from the 5th abdominal segment (Frost, 1942).

Bouché (1977) grouped earthworms into three different ecological types: epigeic species, which dwell in surface litter; endogeic species which are active in mineral soil layers (Haimi and Einbork, 1992); and anecic species, which move vertically between deeper soil layers and the soil surface. Table 4.6 lists some characteristics of these ecological groups. The familiar *Lumbricus terrestris* is an example of an anecic species, building permanent burrows and pulling leaf litter down into them. *L. rubellus*, in contrast, moves horizontally through leaf litter with little involvement in the soil, an epigeic species. Several megascolecid species, such as native earthworms in the southeastern United States, in the genus *Diplocardia*, are endogeic in life habits.

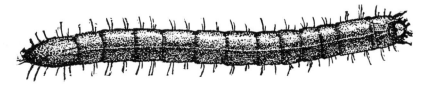

FIGURE 4.35   Larva of an elaterid beetle, the corn wireworm (*Melanotus cribulosis*) (Metcalf and Flint, 1939).

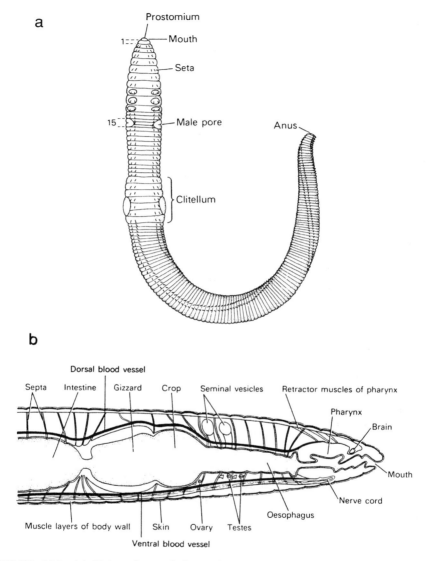

FIGURE 4.36 (a) External morphology of a "typical" lumbricid earthworm. (b) Schematic longitudinal section of the anterior 24 segments of a "typical" lumbricid earthworm, illustrating the disposition of the principal organs of ingestion, digestion, reproduction, blood circulation, locomotion, and coordination (from Lee, 1985).

Some earthworm species appear to be intermediates between these categories, and some do not fit them very well: the manure worm *Eisenia fetida*, for example, or those species inhabiting decaying logs.

TABLE 4.6.   Ecological Groupings of Earthworm Species[a]

Characteristics	Litter species (epigeics)	Topsoil species (anecics)	Subsoil species (endogeics)
Burrows	None	Permanent openings to surface	Extensive, often deep, constantly extended
Casts	No recognizable casts	Cast at surface or in adjacent soil spaces	Cast in burrows or other soil spaces, rarely at surface
Pigmentation	Often heavy; effectively camouflaged	Medium, dorsal only, no ventral pigmentation; little camouflage	Unpigmented or lightly pigmented dorsally; no camouflage
Food	Decomposing litter at soil surface; little or no soil ingested	Decomposing litter collected from soil surface and drawn into burrows; some soil ingested	Much soil ingested with organic matter; may feed on dead roots

[a]Modified from Lee (1985) and Wallace (1994).

Nonetheless, Bouché's ecological categories have become a popular means for segregating earthworm communities into functional groups of species.

Abundance and biomass of earthworms establish them as major factors in soil biology. Biomasses of lumbricid species in temperate regions of the world, where they have been spread by human activities, greatly exceed other animal groups. Lumbricids become dominant in disturbed soils and are particularly numerous in agricultural situations where reduced tillage has been employed (Curry et al., 1995; Edwards and Bohlen, 1995). In the Georgia piedmont, for example, Hendrix et al. (1987) reported an earthworm dry matter biomass of 10 g carbon per m² in no-tillage agricultural plots, a value exceeding all other fauna combined. Earthworm biomass approached the fungal biomass of 16 g carbon per m². Earthworm respiration accounted for 71 g carbon per m² per season. Only bacteria accounted for more respired carbon. In forested ecosystems in England, earthworms consumed more than 90% of the autumnal litterfall in a beech–oak forest (Edwards and Heath, 1963). Earthworms have some enzymes capable of decomposing chitin and oligosaccharides, but the gut milieu itself may promote activities of cellulase and mannanase for the tropical earthworm, *Pontoscolex corethrurus* (Zhang et al., 1993).

Earthworms have pronounced effects on soil structure as a consequence of their burrowing activities as well as their ingestion of soil. Earthworms are responsible for the considerable mixing of the upper soil layers—"bioturbation," an activity where they are rivaled only by ants and termites. Earthworm activity is accompanied by increased numbers of water-stable aggregates: mineral soil particles clumped into rather resistant, larger structures and altered microbial biomass (Daniel and Anderson, 1992). The effect of water-stable aggregates is to increase soil aeration and drainage, factors important in soil fertility (Syers et al., 1979 a,b). The larger burrowing earthworms create biopores, macropores in the soil that allow for downward drainage of soil water and improved aeration, and also reduce runoff (Roth and Joschko, 1991). However, macropores created by earthworms can also allow fertilizers dissolved in ditchwater in an irrigation system to flow below the rooting zone. Fertilizers and other agricultural chemicals may accumulate in ground water, in agroecosystems where earthworms are abundant.

GENERAL ATTRIBUTES OF FAUNA IN SOIL SYSTEMS

Soil scientists and ecologists have begun showing an interest in measuring "soil quality." This elusive concept has been the subject of entire symposia and volumes resulting from them (e.g., Doran et al., 1994; Stork and Eggleton, 1992). As defined by soil scientists, soil quality can be considered as the degree or extent to which a soil can (1) promote biological activity (plant, animal, and microbial), (2) mediate water flow through the environment, and (3) maintain environmental quality by acting as a buffer that assimilates organic wastes and ameliorates contaminants (Linden et al., 1994). Many environmental scientists are attempting to use the concept of indicator organisms or indicator communities as indicators of overall soil "health" (e.g., Bongers, 1990; Ettema and Bongers, 1993; Foissner, 1994; Linden et al., 1994; Neher et al., 1995). Because of their large size and public awareness of them, earthworms are often considered a sign of soil "health" (Linden et al., 1994; Hendrix, 1995). All of the biota play important roles in affecting and influencing soil processes. As summarized in Table 4.7, each of the biotic groups has significant impacts. Among the fauna, microfauna have a principal role via interactions with the microflora. The mesofauna and macrofauna create fecal pellets and produce biopores of various sizes, which affect water movement and storage, as well as root growth and proliferation. Perhaps more importantly, over the longer term, they have marked effects on

TABLE 4.7.   Influences of Soil Biota on Soil Processes in Ecosystems[a]

	Nutrient cycling	*Soil structure*

Microflora	Catabolize organic matter
	Mineralize and immobilize
	  nutrients
		Produce organic compounds that
		  bind aggregates
		Hyphae entangle particles onto
		  aggregates

Microfauna	Regulate bacterial and fungal
	  populations
	Alter nutrient turnover
		May affect aggregate structure
		  through interactions with
		  microflora

Mesofauna	Regulate fungal and microfaunal populations Alter nutrient turnover Fragment plant residues	Produce fecal pellets Create biopores Promote humification
Macrofauna	Fragment plant residues Stimulate microbial activity	Mix organic and mineral particles Redistribute organic matter and microorganisms Create biopores Promote humification Produce fecal pellets

[a]From Hendrix *et al.* (1990)

humification processes as well (Wolters, 1991). Based on biological characteristics, three general trophic systems can be considered: microtrophic (protozoa, nematodes, and some enchytraeids), mesotrophic (the mesofauna), and macrotrophic (the large fauna capable of breaking through physical barriers of soil) (Heal and Dighton, 1985).

Further concerns about fauna as indicators of soil quality led Linden *et al.* (1994) to erect a hierarchical array of three categories, namely: (1) organisms and populations, relating to behavior, physiology, and numbers; (2) communities, with concerns about functional groups, i.e., guilds of burrowers and nonburrowers, trophic groups, and biodiversity; and (3) biological processes, relating to the several processes and products shown earlier (Table 4.8) (Linden *et al.*, 1994). These processes are considered in greater detail in the next chapters on decomposition and nutrient cycling processes.

SUMMARY

Animals in soils are a large, numerous and diverse group of species, organized into complex food webs. The soil fauna may be classified in several ways, in addition to a formal taxonomic classification:

TABLE 4.8.   Properties of Soil Fauna for Use as Indicators of Soil Quality[a]

1. Organisms and populations
    Individuals
        Behavior, morphology, and physiology
    Populations
        Numbers and biomass
        Rates of growth, mortality, and reproduction
        Age distribution
2. Communities
    Functional groups
        Guilds (e.g., burrowers vs nonburrowers, litter vs soil dwellers, etc.)
    Trophic groups
        Food chains and food webs (microbivores, predators, etc.)
    Biodiversity
        Species richness, dominance, and evenness
        Keystone species
3. Biological processes
    Bioaccumulation
        Heavy metals and organic pollutants
    Decomposition
        Fragmentation of organic matter
        Mineralization of C and nutrients
    Soil structure modification
        Burrowing and biopore formation
        Fecal deposition and soil aggregation
        Mixing and redistribution of organic matter

[a]From Linden *et al.* (1994)

persistence in the soil, distribution through the soil profile, body shape, and body size. The latter classification, body size, has the advantages of separating fauna into groups collected and quantified in similar manners. Methods for study of the microfauna, the Protozoa, are essentially the methods of microbiology. Among the mesofauna, the abundant and ubiquitous nematodes have significant impacts on microbial populations and on roots. Another group, the microarthropods, contains mites and collembolans which feed on plant debris rich in fungus, nematodes, and other arthropods as well. The combination of microbes, nematodes, and microarthropods provides complex food webs whose connections may vary opportunistically. The macrofauna contains a large group of arthropods, including the familiar isopods and millipedes as detritus feeders, and scorpions, spiders, and others as predators. Pterygote (winged) insects are numerous in soils. The termites and ants are important soil movers (bioturbators) in many situations, as are earthworms. The earthworms may be the single most important group of soil animals in terms of their feeding on detritus and their effects on soil structure. But the entire fauna is involved in maintenance of soil health. The microfauna and microfloral interactions, the feeding of the mesofauna on microbial-rich detritus, and the creation of biopores and the bioturbation effects of the larger mesofauna all interact in creating soil quality.

5 | *Decomposition and Nutrient Cycling*

INTRODUCTION

The bulk of terrestrial net primary production, along with the bodies and excretions of animals, is returned to the soil as dead organic litter. Some 90% of NPP eventually enters the soil system through dead plants of the grasslands; leaves, roots, and wood from forests; and organic residue from agricultural fields. Indeed, ecosystems may be viewed as consisting of four functional subsystems: the producer subsystem, the consumer subsystem, the decomposition subsystem, and the abiotic subsystem. The decomposition subsystem serves to reduce dead residues to soil organic matter and to release nutrient elements for entry into soil food webs, and ultimately for reaccumulation by plants. The decomposition process drives complex belowground food webs, in which chemical forms of nutrient elements become modified. It is responsible for the creation of long-lived organic compounds important in nutrient dynamics and fuels the creation of soil structure.

Decomposition *per se* is the catabolism of organic compounds in plant litter or other organic detritus. As such, decomposition is mainly the result of microbial activities. Few soil animals have the cellulases which would allow them to digest plant litter. Animal nutrition depends on the action of microbes, either free-living associates in the soil or specialized microbes in animal guts. However, the term "decomposition" is often used as a catchall term to refer to the breakdown or disappearance of organic litter. In that context, the decomposition of organic residue involves the activities of a variety of soil biota, including both microbes and fauna, which interact together. The term "litter breakdown" has been applied to the interactive process which results in the disappearance of organic litter.

INTEGRATING VARIABLES

In studies of soil systems, rates of litter breakdown have been used as integrating variables. That is, since litter breakdown rates are the result of the combined activities of the soil biota, breakdown rates may be used to evaluate effects of disturbance on the entire system. For example, conversion of agricultural systems into conservation tillage regimes will affect soil biology, notably by increases in earthworm populations, but with changes in other soil biota as well. Evaluation of the consequences of these changes in soil biota can be made by measuring litter breakdown rates (Fig. 5.1) (Crossley *et al.* 1992).

Other integrating variables include (1) soil respiration, (2) formation of soil structure, and (3) nutrient dynamics. All of these variables are readily measured and all are important for ecosystem function. Soil respiration estimates biological activity generally and is dominated by microbes, with an important contribution by roots (Cheng *et al.*, 1993). Soil structure is the result of combined actions of biota and climate on mineral substrates. Nutrient dynamics are the most valuable of the integrating variables for predicting performance of crop plants.

Although microbes are responsible for the biochemical degradation of organic litter, fauna are important in conditioning the litter and in aiding microbial actions. The soil scientist Hans Jenny characterized soil fauna as mechanical blenders: "They break up[1] plant material, expose organic surface areas to microbes, move fragments and bacteria-rich excrement around and up and down, and function as homogenizers of soil strata" (Jenny, 1980). Breakdown rates for organic litter integrate the effects of these various activities into a single variable.

---

[1]While we are accustomed to the term "litter break*down*," the term "litter break*up*" appears to be exactly equivalent.

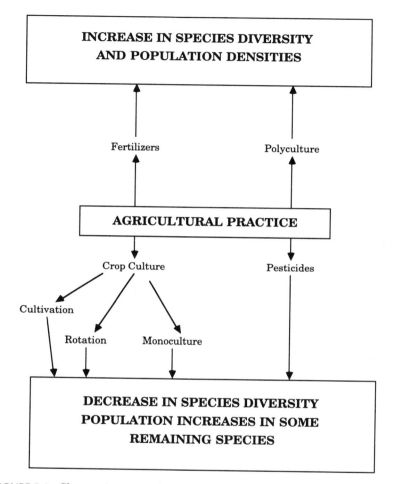

FIGURE 5.1   Changes in species diversity of soil microarthropods as a function of agricultural practice (Crossley *et al.*, 1992).

CLIMATE AND LITTER BREAKDOWN

Litter breakdown rates vary between and among ecosystems on localized and broad geographic scales as functions of soil biota, substrate quality, microclimate, and ecosystem condition. In general, breakdown and decomposition are viewed as the result of biota acting on substrate quality within the constraints of climate.

Organic litter in most terrestrial ecosystems is a mixture of relatively labile and relatively recalcitrant substrates; e.g., thin, calcium-rich dogwood (*Cornus florida*) leaves versus thick, heavy oak leaves (*Quercus* spp.). Even in agricultural systems, differences between leaves and stalks of corn (*Zea mays*), for example, represent different substrate qualities with different breakdown rates. Woody litter may have breakdown rates measured in decades or more. Fine root turnover may be measured in days, but coarse root turnover in years. Microclimate—temperature and moisture around decomposing substrates—regulates activity rates of the biota. On a broader geographic basis, the change in breakdown rates as a function of latitude is generally predictable (Fig. 5.2) (Meentemeyer, 1978).

The effect of latitude on litter breakdown rates is not strictly a direct effect of climate. The abundance of the various soil biota also changes with latitude (Fig. 5.3) (Swift *et al.*, 1979). Adaptations of the soil biota to desert conditions allow breakdown rates to proceed more rapidly than predicted by temperature–moisture considerations. Desert soil biota are active nocturnally, when temperatures are moderate and light dew may accumulate. Litter breakdown in tropical systems may be strongly influenced by the seasonality of litterfall as well as by the faunal abundance. Disturbed ecosystems and successional ones also may have litter breakdown rates that are slower

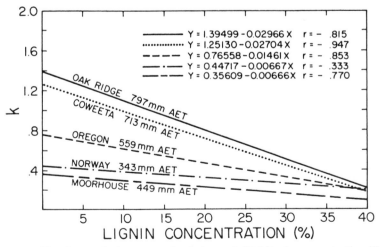

FIGURE 5.2 Simple correlation-regression between initial lignin concentration (%) and annual decomposition rate (K) for five locations ranging in climate from subpolar to warm temperate. AET, actual evapotranspiration (from Meentemeyer, 1978, with permission).

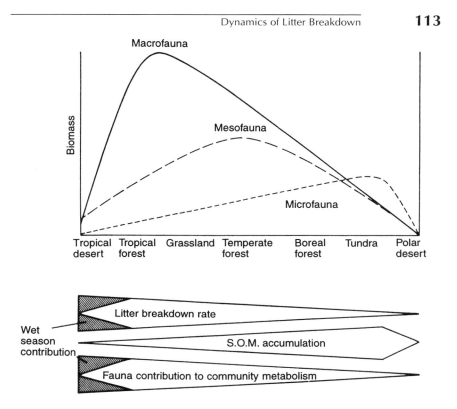

FIGURE 5.3 Hypothetical patterns of latitudinal variation in the contribution of the macro-, meso-, and microfauna to total soil fauna biomass. The effects on litter breakdown rates of changes in the relative importance of the three fauna size groups are represented as a gradient together with the faunal contribution to soil community metabolism. The favorability of the soil environment for microbial decomposition is represented by the cline of soil organic matter accumulation from the poles to the equator; SOM accumulation is promoted by low temperatures and waterlogging where microbial activity is impeded (from Swift *et al.*, 1979).

than predicted from broad regional temperature–moisture conditions. Alteration of microclimates may reduce faunal activities, and substrate quality of foliage may change during plant succession.

DYNAMICS OF LITTER BREAKDOWN

The disappearance of litter on forest floors follows approximately a simple first-order equation

$$dX/dt = -kX,$$

where X is the standing crop of litter and k is the annual fractional rate of disappearance. Olson (1963) proposed that it was a characteristic of mature forests that rates of litter production and disappearance were equal so that annual production (L) would be balanced by breakdown ($-kX$). Olson used the symbol X_{SS} to designate the standing crop of litter at steady state (litter production and disappearance equal). Then, the ratio of input (L) to standing crop (X_{SS}) provides an estimate of breakdown rate k:

$$k = L/X_{SS}.$$

Olson (1963) estimated decomposition rates (k) for evergreen forests in various parts of the world (Fig. 5.4). Values for k ranged from 4 for rapid decomposition in tropical regions, through 0.25 for eastern United States pine forests, to 0.02 for higher latitude pine forests.

What is estimated here is the rate of leaf or needle litter breakdown. Olson (1963) did not consider inputs of organic litter belowground, although he did use the entire mass of carbon per square meter in estimates of X_{SS}. Current studies of root dynamics (see Chapter 2) provide estimates of root breakdown rates, but these are more difficult to measure than leaf litter breakdown rates and consequently are less well known. Root death and decay may account for as much as one-half of the annual carbon addition to forests, but as is often the case, dynamics within the soil are obscure.

The simple exponential model using a single constant, k, to represent the decomposition rate continues to be widely used. It is not difficult to estimate k using litterbag techniques (see later). The simple model loses its attractiveness when patterns of litter breakdown are examined more closely. Leaf litter often is a combination of leaf species, each with different breakdown rates. Furthermore, each species contains both labile and recalcitrant fractions. Wieder and Lang (1982) examined several different models and concluded that the single exponential model shown earlier or double exponential models (including fast and slow components) best describe breakdown rates over time "with an element of biological realism."

DIRECT MEASUREMENT OF LITTER BREAKDOWN

In deciduous forests with annual leaf drop, it is possible to measure litter breakdown directly from samples taken through time. If combined with estimates of the mass of litterfall, these samples provide a good measure of the dynamics of litter breakdown (Fig. 5.5) (Witkamp and van der Drift, 1961). As the year progresses the litter layer mass becomes transformed into an F layer (see Chapter 1). Sampling over

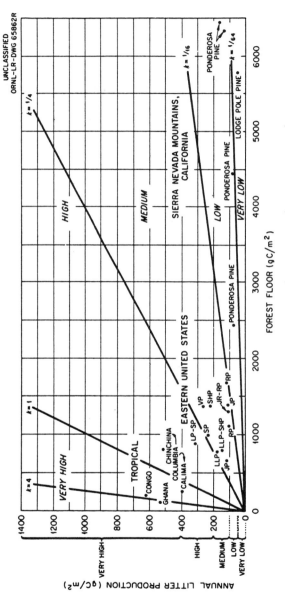

FIGURE 5.4 Estimates of decomposition rate factor k for carbon in evergreen forests, from the ratio of annual litter production L to (approximately) steady-state accumulation of forest floor X_{ss} (from Olson, 1963, with permission).

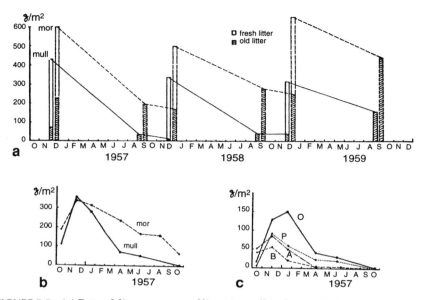

FIGURE 5.5   (a) Rate of disappearance of litter in mull and mor. (b) Amount of litter on a cleaned surface in mull and mor. (c) Amount of fresh oak (O), birch (B), poplar (P), and alder litter (A) in mull (from Witkamp and van der Drift, 1961).

several years reveals a year-to-year variation in masses of litter input and rates of breakdown (Table 5.1).

Rates of litter breakdown are measured more easily using bagged leaf litter. Mesh bags (litterbags) containing a known mass of leaf litter are placed on the forest floor at the time of leaf drop. Litterbags are then collected on a time schedule, and the remaining mass is measured (Fig. 5.6). Litterbags have been a valuable tool for comparative studies of rates of litter breakdown (Fig. 5.7). Such studies include mass loss rates by different tree species and have shown the importance of elemental contents, lignin, C/N ratios, and other resource quality factors (Table 5.2) (Melillo *et al.*, 1982). Decomposition rates also vary between forest types, and litterbags have proved to be useful in delineating and analyzing differences (Table 5.3) (Cromack, 1973).

Use of litterbags does have its problems. Fine mesh bags, with openings of 1–2 mm, will exclude most macrofauna and thus underestimate decomposition rates. Larger meshes allow larger fragments to escape the bags, thus overestimating decomposition. The microclimate within litterbags tends to be more moist than that of unbagged litter, and thus more favorable for microbial activity (Vossbrinck *et al.*,

TABLE 5.1. Summary of Litter Decomposition Experiments Conducted on a Clearcut (WS 7) and Control (WS 2) Watershed at the Coweeta Hydrologic Laboratory[a]

Species	1974–1975 WS 7, precut		1975–1977 WS 2, precut		WS 7, postcut		1977–1978 WS 2, postcut	
	Rate	% Remaining	Rate	% Remaining	Rate	% Remaining	Rate	% Remaining
Liriodendron tulipifera	−0.682	49.4	−0.656	49.8	−0.545	60.0	−0.814	47.3
Acer rubrum	−0.529	49.0	−0.477	57.9	−0.324	71.5	−0.368	67.9
Quercus prinus	−0.336	69.3	−0.285	72.2	−0.242	79.3	−0.300	76.0
Cornus florida	−1.309	27.8	−0.711	47.8	−0.531	59.6	−0.825	43.4
Robinia pseudoacacia	−0.250	72.4	−0.530	49.7	−0.330	69.1	−0.330	70.7

Note. Data are summarized for a 1-year study on WS 7 (1974–1975), a 2-year study on WS 2 (1975–1977), and a 1-year study on WS 2 and 7 (1977–1978). Values are shown for decay rate (per year) and percent remaining after 1 year. Only first-year decay results are presented here.
[a]From D. A. Crossley, Jr., unpublished.

FIGURE 5.6   View of leaf litterbag.

TABLE 5.2.   Initial Litter Quality Variables as Predictors of First-Year Decay Rates[a]

Initial litter quality variable	r^2	Slope	Y intercept
% nitrogen	0.271	−0.978	−1.29
C:N ratio	0.138	0.007	−0.12
% lignin	0.987	0.029	−0.96
Lignin:N ratio	0.967	0.027	−1.05
% water soluble	0.322	−0.015	−0.06
% ethanol soluble	0.426	−0.040	−0.32

Note. Coefficients of determination (r^2), slopes, and Y intercepts of regressions relating first-year decay rate constants (k) to initial litter quality variables for litter of the three species examined.

[a]Reprinted from Blair, J. M. (1988a). Nitrogen, sulfur, and phosphorus dynamics in decomposing deciduous leaf litter in the southern Appalachians. *Soil Biol. Biochem.* **20**, 693–701. Copyright 1988, with kind permission from Elsevier Science Ltd, The Boulevard, Langford Lane, Kidlington 0X5 1GB, UK.

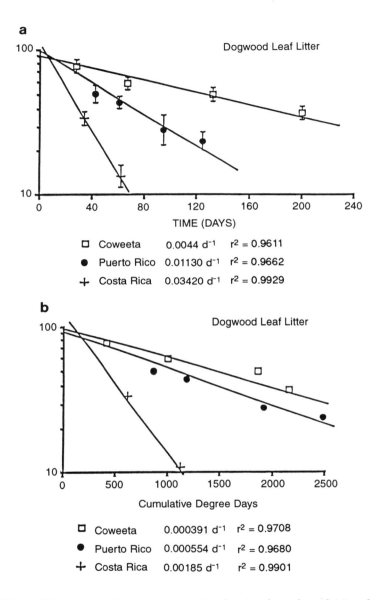

FIGURE 5.7 Litterbags and masses remaining showing days elapsed (a) and degree days (b) at one temperate (Coweeta) and two tropical (Puerto Rico and Costa Rica) sites (Crossley and Haines unpublished, 1990).

TABLE 5.3. First-Year Litter Breakdown Rates for Single Species[a]

Species	Year	% Weight remaining	Exponential loss rate (k^b)	Correlation coefficient (r)
White pine	1969–1970	59.5	−0.52 (12)	−0.888**
	1970–1971	65.4	−0.42 (19)	−0.480*
Chestnut oak	1969–1970	57.8	−0.55 (33)	−0.866**
	1970–1971	51.4	−0.66 (31)	−0.890**
White oak	1969–1970	47.6	−0.74 (36)	−0.887**
	1970–1971	49.6	−0.70 (32)	−0.906**
Red maple	1969–1970	43.6	−0.83 (36)	−0.934**
	1970–1971	48.6	−0.72 (30)	−0.839**
Dogwood	1969–1970	30.8	−1.18 (37)	−0.939**
	1970–1971	26.0	−1.35 (33)	−0.948**

[a]From Cromack (1973).
  * $P<0.05$.
 ** $P<0.01$.
[b]Where the annual exponential loss rate (k) is estimated from a semilogarithmic regression (base e) of monthly weight loss of litterbags.

1979). In most cases, litterbags probably underestimate actual breakdown rates. Their usefulness for comparative studies and for nutrient measurements makes them important tools nevertheless.

As an alternative to litterbags, individual leaves tied together by their petioles on a string ("trot lines") have been used. Loss of weight (and area) by individual leaves measured through time yields estimates of litter breakdown rates. When biological activity increases in late spring and summer, rapid rates of loss are found. It is not clear whether these rapid losses are due to the separation of large fragments from the leaf, whether the unbagged rates allow for larger fauna to attack the decomposing leaf, or both. The simultaneous use of both techniques yields estimates of breakdown rates which doubtless bracket the true values.

PATTERNS OF MASS LOSS DURING DECOMPOSITION

A graph of mass retained in litterbags during the decomposition of forest leaf litter reveals a three-phase curve (Fig. 5.8). Initially, following autumnal leaf drop, there is a rapid decrease in weight due to the loss of rapidly metabolizable compounds or simply readily leachable substances. This initial phase is followed by a slow rate of loss during winter months. During late spring, rates again become accelerated as microclimates become more favorable for biological

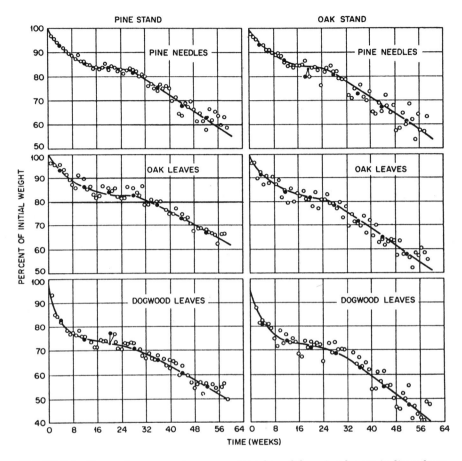

FIGURE 5.8 Three-phase mass loss curve. Weights of decaying leaves in litter bags, expressed as a percentage of initial weight of litter, through time. Three leaf species in two stands. Hollow circles are individual measurements; solid circles are averages for 8-week cycles. Lines fitted by eye (from Olson and Crossley, 1963).

activity. During winter some microbial and faunal attack occurs, but the major abundance of fauna and microbes is found in litterbags during spring and summer.

Although rates do vary with season, the model using a single exponential constant (k) provides a good fit to these data (Fig. 5.9). The coefficient of determination (r^2) for these curves usually exceeds 85%. The constant k conceals seasonal dynamics but is a useful means for

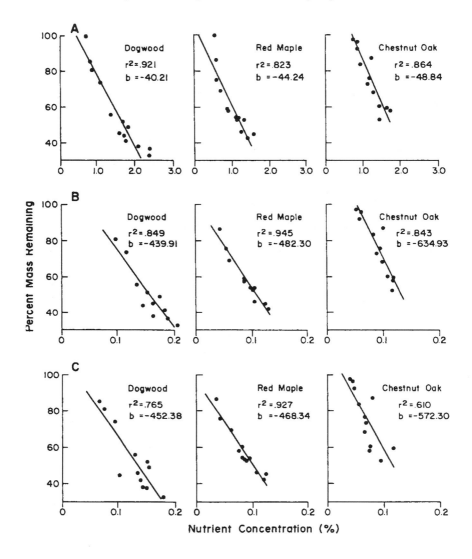

FIGURE 5.9 Single K value for decay [regressions of % mass remaining over (A) nitrogen, (B) sulfur, and (C) phosphorus concentrations in the residual litter for flowering dogwood, red maple, and chestnut oak]. (Reprinted from Blair, J. M. (1988a). Nitrogen, sulfur and phosphorus dynamics in decomposing deciduous leaf litter in the southern Appalachians. *Soil Biol. Biochem.* **20**, 693–701. Copyright 1988, with kind permission from Elsevier Science Ltd. The Boulevard, Langford Lane, Kidlington 0X5 1GB, UK.)

comparing leaf types, habitats, or geographical regions. More precision can be gained by calculating k from the spring–summer values alone. Some typical breakdown rates for forest litter are shown in Table 5.3.

Breakdown rates in agricultural systems are generally more rapid than in forested systems: crop residues, as a rule, tend to have fewer recalcitrant components. Figure 5.10 shows mass loss rates for rye litter from litterbags either placed on the soil surface (no-tillage) or buried following plowing (conventional tillage) (Beare *et al.*, 1992).

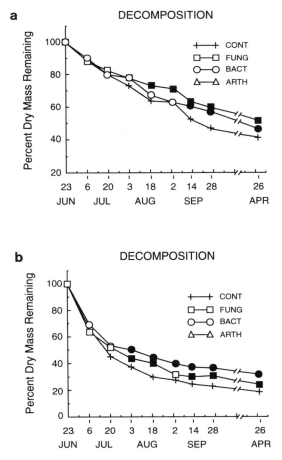

FIGURE 5.10   Mass loss rates for surface (a) and buried (b) rye litter over 320 days. CONT, Control situation; FUNG, fungicide, with ca. one half fungal population biomass; BACT, bactericide (oxycarbon); ARTH, arthropod repellant (naphthalene). (Modified from Beare *et al.*, 1992, with permission.)

Loss rates were much faster under conventional tillage ($k = 0.03$ per day) than in no-tillage soils ($k = 0.02$ per day). Usually, buried residues decompose more rapidly than surface residues. In any case, rates for rye litter were much faster than for forest tree leaf litter.

EFFECTS OF FAUNA ON LITTER BREAKDOWN RATES

The Russian soil scientist Galina Kurcheva (1960, 1964) found that naphthalene (an insecticide) applied to oak leaf litter drastically reduces the rate of breakdown. In the succeeding decades, various biocides and other techniques have been used to suppress various elements of the soil biota as a measure of their importance in leaf litter breakdown. The upshot of these experimental manipulations has been to demonstrate that bacterial, fungal, and faunal members of the soil biota all have significant effects on litter breakdown (see Fig. 5.10). Given that actual breakdown and decomposition rates are a product of the *interaction* of the various biota (and substrate quality and climate), the rate estimates derived from manipulations must be accepted with caution. The main effects, however, seem clear.

Seastedt (1984b) suggested that the equation describing litter breakdown might be subdivided into components so that the constant k could be considered as the sum of several ks:

$$dX/dt = -kX = -(k_{bacteria} + k_{fungi} + k_{fauna})X .$$

Seastedt reviewed studies in which microarthropods had been suppressed and found that a variable percentage of breakdown rates could be attributed to microarthropod activities. Table 5.4 shows the results of the Seastedt equation applied to forest tree litter in a floodplain forest in Athens, Georgia. Litterbags with a 1-mm mesh size were used so that macrofauna were excluded from the bags. Naphthalene applications were used to reduce microarthropod populations in some of the litterbags. The results show that the importance of microarthropods varied with litter quality. Microarthropod activities were least significant for the more rapidly decomposing litter species (dogwood, tulip-poplar) and were most important for the slowest, most recalcitrant litter type (water oak).

Experimental approaches such as these must be interpreted with caution. Usually more than one element of the system is modified by manipulations, be they chemical or physical ones. Biocides such as naphthalene may alter other system components, sometimes to a large extent. Other approaches, such as tracer methods and laboratory microcosms, need to be used in conjunction with manipulative experiments.

TABLE 5.4. Percentage of Leaf Litter Decomposition (Mass Loss) Attributable to Soil Fauna[a]

Leaf species	k_T	k_{NA}	k_F	Percent due to fauna
Dogwood	−0.00248	−0.00089	−0.00159	64.1
Sweetgum	−0.00248	−0.00089	−0.00159	71.4
Tulip-poplar	−0.00229	−0.00113	−0.00116	50.7
Red maple	−0.00125	−0.00069	−0.00056	44.8
Water oak	−0.00174	−0.00037	−0.00137	78.7
White oak	−0.00216	−0.00076	−0.00140	64.8

Note. Loss rate due to faunal activities calculates as total rate (k_t) minus naphthalene rate (k_{NA}) equals rate due to fauna (k_F). Percent difference calculated as (k_F/k_t) $\times$ 100.

[a]From D. A. Crossley, Jr., unpublished, 1988.

NUTRIENT MOVEMENT DURING DECOMPOSITION

Soil contains the same elements as found in its underlying substrate of rock, but the proportions differ greatly. Elements such as Na, Ca, Mg, and K are lost as soluble cations during weathering. Some other elements, such as Fe and Al, are resistant to leaching losses and their proportions increase compared to rocks. Movements of cations are governed by the exchange properties of the soil, properties dependent on the nature of the clays and the amount and type of organic matter. The most common exchangeable cation in soils is usually calcium. In contrast, K and NH_4^+ may be tightly bound to clay particles. Certain anions are not as tightly held in soils. Phosphate ions, multiply charged, are more tightly fixed by exchange properties than are singly charged ions such as nitrate (Bowen, 1979).

During the decomposition process, elements are converted from organic to inorganic forms (mineralized) and may enter the exchangeable pools from which they are available for plant uptake or microbial use. Cellulose and hemicellulose account for more than 50% of carbon in plant debris and thus fuel microbial processes. Transformations of nitrogen (Fig. 5.11) and sulfur (Fig. 5.12) are microbially driven processes, which gradually reduce the C:N and C:S ratios in decomposing materials.

As plant litter decomposes, the elemental mix changes because of differential mobility and biological fixation. Carbon is lost through microbial respiration as cellulose is hydrolyzed and utilized. Potassium is highly mobile until it encounters clay particles, where it becomes tightly fixed. Sodium, a mobile ion in soils, is not accumulated in plants but is essential for animals. The "herbivore

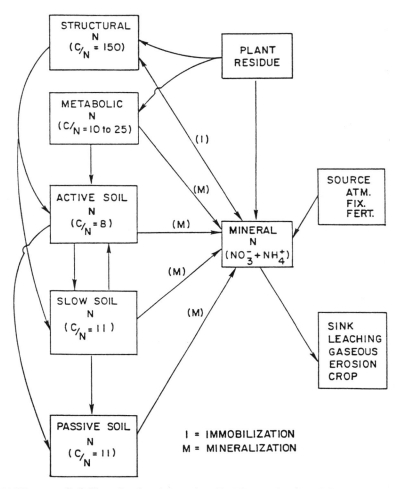

FIGURE 5.11   Soil N cycle showing active (1–1.5 years), slow (10–100 years), and passive (100–1,000 years) fractions. Flow diagram for the N submodel of the century model (from Parton *et al.*, 1987, with permission).

exclusion hypothesis" proposes that plants discriminate against sodium and thereby limit herbivory. Sodium does accumulate in food chains, often increasing by a factor of 2–3 between trophic transfers. Calcium is lost from decomposing litter at about the same rate as mass is lost.

The nitrogen content of decomposing litter increases during the initial stages of decomposition and then declines (Fig. 5.13) (Berg and Staaf, 1981). Nitrogen is mineralized during decomposition and is

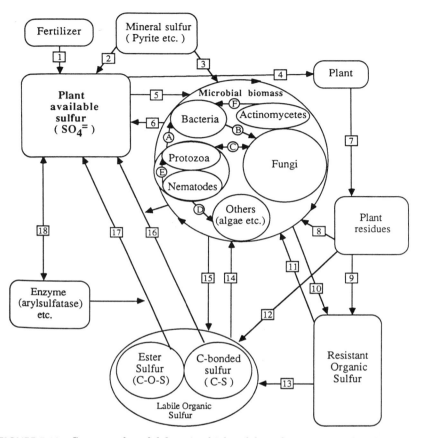

FIGURE 5.12 Conceptual model for microbial and faunal component of sulfur cycle in soil (from Gupta, 1989).

simultaneously immobilized by microbes, resulting in an increase in the absolute amount of nitrogen in the litter. As decomposition proceeds, the C:N ratio declines until the substrate becomes more suitable for microbial action. In some forests the period of nitrogen increase may extend for 2 years or more (Fig. 5.14) (Blair and Crossley, 1988). Phosphorus and sulfur also show increases in absolute amounts during the decomposition of some species of tree leaf litter (Fig. 5.15) (Blair, 1988a), even though mass is being lost. Calcium and magnesium concentrations in decomposing litter change only slightly through time. There may be an initial decrease in concentration followed by a slight increase (Blair, 1988b). Thus, the absolute amounts of these elements during decomposition

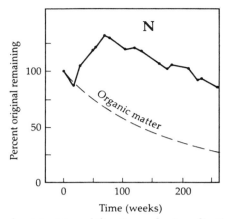

FIGURE 5.13   Rates of nutrient immobilization and mineralization from decomposing Scots pine litter over a 5-year period (from Staaff and Berg, 1982). Note initial influx of N into litter.

approximately track the loss of mass. Potassium is not a structural element and is lost more rapidly than is mass of decomposing leaf litter. Decomposing woody litter, in contrast, accumulates calcium and phosphorus, evidently a result of fungal invasion and translocation.

The nitrogen pool in decomposing litter is a dynamic one. Although nitrogen is accumulating, there is evidently a large amount of turnover taking place. When tracer amounts of ^{15}N were added to leaf litter, significant losses of tracer took place even as total nitrogen accumulated (Fig. 5.16) (Blair *et al.*, 1992). Nitrogen is evidently becoming incorporated from exogenous sources in amounts greater than those lost through biotic factors. Inputs of nitrogen via rainfall or canopy throughfall are a potential source of added nitrogen. However, these would appear to be inadequate to account for the amount of nitrogen immobilized in litter. Fungal translocation from lower layers is another possibility. Finally, lateral transport to and from "hot spots" in the forest floor may contribute to the dilution of tracers.

ROLE OF SOIL FAUNA IN ORGANIC MATTER DYNAMICS AND IN NUTRIENT TURNOVER

For the last several decades there has been interest in the role of soil fauna in litter and organic matter turnover in ecosystems. The pioneering studies of Darwin (1881) and P. E. Müller (1887)

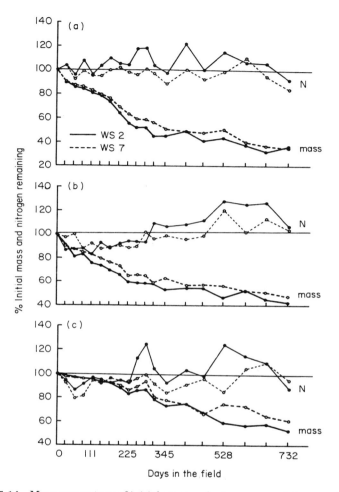

FIGURE 5.14   Mean percentage of initial mass and nitrogen remaining over time in (a) *Cornus florida,* (b) *Acer rubrum,* and (c) *Quercus prinus* litter on uncut WS 2 (—) and clearcut WS 7 (---) from January 1975 to January 1977 (from Blair and Crossley, 1988).

emphasized the prominent signs left in many temperate forest and grassland communities by earthworm, mesofaunal, and biotic activities in general. A very prescient account of the "biotic" structure of soils was given by Jacot (1936). Signs of faunal activity include coating of mineral grains, which has a significant effect on promoting the formation of aggregates (Kubiëna, 1938). Termites in semitropical and tropical regions have similar functions as well (e.g., Lee and Wood, 1971; Wood *et al.,* 1983). Only a few ecologists are aware, however,

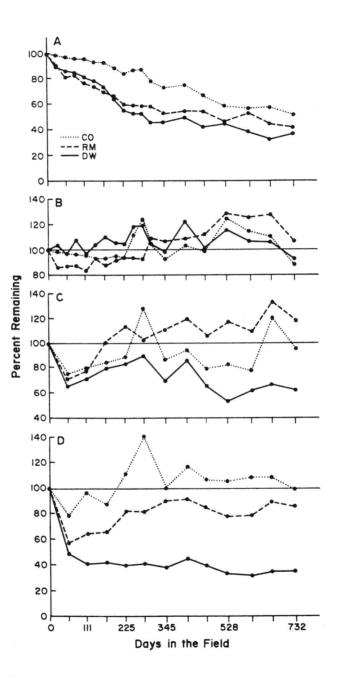

that often the soil meso- and microfauna are vastly more numerous and are usually more active than the large soil fauna (Wolters, 1991).

Our concerns as ecosystem researchers should include both an understanding of what organisms are present and the major processes which they carry out in a wide range of terrestrial ecosystems. Following the flow of energy and nutrients in the system (as noted by Volobuev, 1964) will enable us to concentrate on key processes which occur, avoiding the pitfall of what is obvious to the naked eye being singled out for study. We must get to the appropriate level of resolution to ascertain the roles of participants in soil processes (Macfadyen, 1969; Coleman, 1985). This requires exploring the myriad of surfaces and volumes which occur in a few cubic millimeters of soil and organic matter (Elliott, 1986).

Fauna are members of the "organism" category in the Jenny (1941) factors of soil formation S.F.F. = f (cl,o,p,r,t), where cl is climate, o is organisms, p is parent material, r is relief, and t is time (see Fig. 1.5). As noted by Crocker (1952), only a few of these factors are independent variables, so we are dealing with a multiple-causation, interdependent subset of a terrestrial ecosystem.

To simplify matters, let us consider organisms alone. These are vegetation, organic matter inputs therefrom, and the array of heterotrophs feeding on organic detritus or the organisms decomposing it. The factors plus ecosystem processes acting over time lead to ecosystem properties (Elliott, 1994).

The immediate result of faunal feeding activity is the production of fecal pellets, some of which can be identified as species or group specific (Kühnelt, 1958; Jongerius, 1964; Zachariae, 1965; Rusek, 1975; FitzPatrick, 1984; Pawluk, 1987). A comprehensive review (Bal, 1982) of soil fauna activities in soil refers to "zoological ripening" as faunal movement of organic matter and mineral materials in previously uncolonized soil. This soil maturation and development process has been of great significance in Dutch polder regions and has been demonstrated in Canadian (Nielson and Hole, 1964) and New Zealand (Stockdill, 1966) soils as well. These processes are extensively reviewed in Brussaard and Kooistra (1993).

FIGURE 5.15   Mass N, P, and S. Changes in (A) mass and absolute amounts of (B) nitrogen, (C) phosphorus, and (D) sulfur in flowering dogwood, red maple and chestnut oak litter decomposing over a 2-year period. (Reprinted from Blair, J. M. (1988a). Nitrogen, sulfur and phosphorus dynamics in decomposing deciduous leaf litter in the southern Appalachians. *Soil Biol. Biochem.* **20**, 693–701. Copyright 1988, with kind permission from Elsevier Science Ltd, The Boulevard, Langford Lane, Kidlington OX5 1GB, UK.)

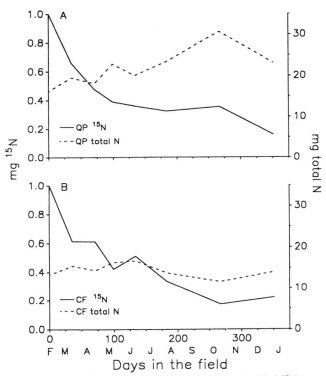

FIGURE 5.16   [15]N and Total N, Changes in the amount of added ^{15}N recovered in the litter (solid line) vs changes in the total amount of N (dashed line) over time in (A) *Quercus prinus* (QP) and (B) *Cornus florida* (CF) litter from control litter baskets. [Reprinted, with permission, from Blair, J. M., Crossley, Jr., D. A., and Callaham, L. C. (1992). Effects of litter quality and microarthropods on N dynamics and retention of exogenous ^{15}N in decomposing litter. *Biol. Fertil. Soils* **12**, 241–252. Copyright 1992 Springer-Verlag.]

In addition to physical signs, there are chemical indicators of faunal presence and activity. For example, in certain cool, moist New Zealand tussock grassland soils, nearly 10% of the organic P is composed of phosphonates (C-P bonded) (Newman and Tate, 1980; Tate and Newman, 1982) as contrasted with the more prevalent phosphate esters. Phosphonates are produced by ciliates, and their subsequent rates of input and flow through the soil P cycle remain unknown (Stewart and McKercher, 1982).

Several authors have reviewed work on experimental pedogenesis (soil formation), examining roles of primary colonizing plants, including dissolution of rock minerals by lichens, as well as faunal impacts on mineral or soil movement and organic matter transformation (Hallsworth and Crawford, 1965; Bal, 1982). Webb (1977) studied the effects of particle size and decomposability of macrofaunal and micro-faunal fecal pellets. There are differing effects of comminution (breaking up) of leaf litter by large and small fauna and they play different roles in facilitating further leaf litter decomposition. Webb (1977) noted that fecal pellets of *Narceus annularis* (Diplopoda: Spirobolidae) had a lower surface-to-mass ratio than the original deciduous leaf litter, whereas microarthropods such as oribatids have a greater surface-to-volume ratio in their fecal pellets compared with the original leaf litter, which should lead to greater decomposition per unit time (Fig. 5.17).

Physical interpretation of organic matter decomposition should be tempered with careful observation of life history details, such as the likelihood of localized aggregation of mite or collembolan fecal pellets which may decompose locally at a much slower rate than hypothesized from *in vitro* laboratory studies. Substrate quality plays an important role here. Dunger (1983) noted that macroarthropods ingest mineral soil along with litter material. Kilbertus and Vannier (1981) and Touchot *et al.* (1983) demonstrated ingestion of argillic (clay) material by *Tomocerus* and *Folsomia* sp. (Collembola), a trait that was

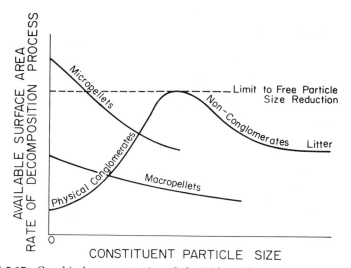

FIGURE 5.17 Graphical representation of physical conglomerate feces differentiation theory (Webb, 1977). As particle size of litter (right to left) is reduced, surface area and decomposition increase until constituent particles are small enough to aggregate into more stable conglomerates (limit to free particle size reduction). Physical conglomerates increase in size as constituent particle size decreases, but arthropod pellets decrease in size due to the direct relationship of body size to degree of pulverization and pellet (conglomerate) size. Micropellets are therefore able to maintain a much smaller conglomerate size and break the limit to free particle size reduction (from Webb, 1977).

particularly evident when they ingested polyphenol-rich *Quercus* leaves. This detoxification process presumably led to greater decomposition of the leaf material, with enhanced bacterial growth in the pellets with clay particles versus those without the clay adsorbent material. The impact of collembola is greatest in mor soils, which may have entire layers in the F or H horizon filled with collembolan fecal pellets (Pawluk, 1987).

FAUNA AND NUTRIENT CYCLING

Research since the mid-1980s has shown a significant impact of root-associated organisms on nutrient dynamics of phosphorus and nitrogen in experimental microcosms. These studies are reviewed in Coleman *et al.* (1983) and Anderson *et al.* (1981a,b). We now discuss results from our laboratory in microcosm (Ingham *et al.*, 1985) and field experiments (Ingham, 1986a,b, 1989; Parmelee *et al.*, 1990; Beare *et al.*, 1992).

In the laboratory, groups of rhizosphere bacteria, fungi, and microbivorous nematodes were grown singly or in combination, all with growing seedlings of the shortgrass prairie grass *Bouteloua gracilis* (blue grama). In all treatments which had the root, microbe, and microbial grazer (*Pelodera* sp. as bacterial-feeder) and *Aphelenchus avenae* as the fungal feeder, there was an enhanced shoot growth and dry matter yield when compared to the plant alone control.

Other work using mesofauna (nematodes; Ingham *et al.*, 1986a,b) and macrofauna (isopods, Anderson *et al.*, 1985) has shown a significant enhancement of nutrient cycling (nitrogenous compounds) in field experimental situations. Thus an enhanced (20–50%) nutrient return (mineralization) occurs in the presence of the fauna compared with experiments in which they are present in very low numbers or completely absent (Anderson *et al.*, 1983). This work was further amplified by simulation models of detrital food webs, which showed a significant (ca. 35%) contribution to mineralization of nitrogen by microfauna (amoebae and flagellates) and bacterial-feeding nematodes (Hunt *et al.*, 1987; De Ruiter *et al.*, 1993).

FAUNAL IMPACTS IN APPLIED ECOLOGY: AGROECOSYSTEMS

There are several areas in the interface between theoretical and applied ecology where our knowledge of soil physics, chemistry, and biology can, and should, be put to good use. One of these is in the area of agroecosystem studies. The essentials of decomposition and nutrient dynamics in agroecosystems have been reviewed (Hendrix *et al.*, 1992; Coleman *et al.*, 1993).

It is generally acknowledged that zero, or reduced, tillage has several effects on abiotic and biotic regimes in agroecosystems. Retention of litter keeps the surface of the soil cooler and moister than in a conventionally tilled plot (Fenster and Peterson, 1979; Phillips and Phillips, 1984) and also leaves more substrate available in the 0 to 7.5-cm depths for nitrifiers and denitrifiers (Doran 1980a,b). This abiotic buffering seems to promote a slower N cycle, one which continues over a longer time span, but at a lower rate per unit time (House *et al.*, 1984; Elliott *et al.*, 1984). Soil invertebrate populations, particularly microarthropods (Stinner and Crossley, 1980; House *et al.*, 1984) and earthworms, are enhanced as well (Table 5.5) (Coleman and Hendrix, 1988). The microarthropods are undoubtedly responding to increased populations of litter-decomposing fungi, which tend to concentrate nitrogen by hyphal translocation (Holland and Coleman, 1987). In fact, dominant families of fungivorous mites responded by markedly decreasing their numbers in field mesocosm plots which were treated with captan to keep fungal populations down to ca. 40% of normal levels (Mueller *et al.*, 1990).

A total of 22 agroecosystem components and processes were compared in no-till and conventional tillage in Georgia. In many instances, there was greater resilience in the system, as shown by greater invertebrate species richness, greater soil organic matter, and ecosystem N turnover time (Table 5.6) (House *et al.*, 1984).

These findings were confirmed and extended by Elliott *et al.* (1984), who examined dynamics in long-term stubble mulch and no-till plots on a silty-loam soil in eastern Colorado which underwent alternate crop and fallow regimes. These plots have been under cultivation since 1920, and no-till has been an experimental treatment since the mid-1970s. Nitrate accumulated to a greater extent in the fallow than in the cropped rotation (Table 5.7). Ammonium-N was usually at very low levels (~1.0 µg NH_4^+ - N·g^{-1} soil), but on one date the concentration reached 4.6 µg NH_4^+ - N·g^{-1} soil in the top 2.5 cm of the no-till plots just prior to the highest rate of NO_3^- accumulated in the no-till than in the stubble mulch treatments. However, it is possible that there was more mineralization in the stubble mulch plots earlier in the year before the first sample date, and this mineralized N was moved below the sampling depth (20 cm) as NO_3^- - N during a rainfall event. Interactions between modification of system structure and major nutrient processes need more study. Certainly soil fauna are sensitive to increased nutrient inputs from fertilizers and manures, and this needs to be considered in experimental work (Marshall, 1977).

TABLE 5.5.   Numbers and Estimated Biomass of Soil Fauna in Conventional
Tillage (CT) and No-Tillage (NT) Agroecosystems at Horseshoe Bend[a]

	Numbers*m^{-2}		mg dry wt*m^{-2}	
	CT	NT	CT	NT
Nematodes [b]				
Bacterivores	1,836	909[f]	237	117
Fungivores	227	500[f]	14	31
Herbivores	945	1,064	93	104
	3,008	2,473	344	252
Microarthropods [c]				
Mites	41,081	78,256[f]	118	303
Collembola	6,244	14,684[f]	17	40
Insects	2,105	2,548	—	—
	49,430	95,488	135	343
Macroarthropods [d]				
Ground beetles	7	33[f]	6	30
Spiders	1	17[f]	1	14
Others	6	28[f]	—	—
	14	78	7	44
Annelids [e]				
Earthworms	149	967[f]	3,129	20,307
Enchytraeids	1,867	520	592	17
	1,986	1,487	3,188	20,963
Total	54,438	99,526	3,674	20,963

[a]From Hendrix *et al.* (1986), with permission. © 1986 American Institute of
Biological Sciences.
   [b]Means of samples from June to October 1983; numbers are × 10^{-3}.
   [c]Means of samples from May to December 1983.
   [d]Means of samples from April to June 1983.
   [e]Means of samples from April 1983.
   [f]For numbers of organisms, tillage treatments differ significantly at $P = 0.05$.

APPLIED ECOLOGY IN FORESTED ECOSYSTEMS

Some interesting comparisons and analogies can be drawn between
no-till agriculture and forested ecosystems of the "mor" type, which
have a distinct stratification of L, F, and H (A$_o$, A$_i$, and A$_{ii}$ in the North
American terminology). It is generally recognized that abundance of
fungi and fungivorous arthropods is greater in these soils than in soils
with a less pronounced litter layer (Kühnelt, 1976; Wallwork, 1976;
Pawluk, 1987; Blair *et al.*, 1992). However, it is important to determine the amount of activity occurring in these surface layers as well.
Ingham *et al.* (1989) and Coleman *et al.* (1990) have shown significantly greater fungal biomass and microarthropod biomass in L, F,
and H layers under *Pinus contorta* (lodgepole pine) compared with

TABLE 5.6. Comparison of Agroecosystem Components and Associated Agroecosystem Processes from Conventional Tillage (CT) and No-Tillage (NT) Systems[a]

Component or process	CT versus NT
Crop yields	NT - CT (except during drought)
Crop biomass	Decreasing in both CT and NT
Weed biomass	NT > CT
Plant nitrogen dynamics	CT > NT (nitrogen flux)
Shoot to root ratios	CT > NT
Nitrogen fixation	NT > CT (?)
Surface crop and weed residues	NT >> CT
Litter decomposition rates	CT > NT
Surface litter (%N)	NT > CT
Soil total N	NT > CT in upper soil layer
Nitrification activity	NT > CT in upper soil layer
	CT > NT in middle soil layer (?)
Soil organic matter	NT > CT
Soil moisture	NT > CT
Groundwater leaching (nitrate-N)	CT > NT (?)
Foliage arthropods	CT - NT
Crop herbivory by insects	CT > NT
Nitrogen content of crop foliage	CT > NT
Arthropods species diversity	NT > CT
Soil arthropods (No. of individuals)	BT >> CT
Nitrogen contained in arthropods	
Soil	NT > CT
Foliage	CT > NT
Ecosystem N turnover time	NT > CT
Ecosystem N efficiency	NT > CT (?)

[a]From House et al. (1984).

mountain meadow. There was also a greater amount of fungal activity, as demonstrated by FDA-positive fungal hyphae (Söderström, 1977; Ingham and Klein, 1982). This forest experience is corroborated by Verhoef and De Goede (1985), who noted greater activity of Collembola in pine forests in Holland contrasted with habitats which had a thin or nonexistent litter layer.

Other areas of interest in applied ecology include revegetation of mine spoils. Extensive studies in the United Kingdom, Germany, and elsewhere have been made of decomposition ecology and microbial parameters in strip-mined coal lands (Bentham et al., 1992).

Several researchers (Tisdall and Oades, 1982; Rothwell, 1984; Jastrow and Miller, 1991) have investigated the roles of saprophytic and VAM fungi in stabilizing macroaggregates. Rothwell (1984) suggests that a biochemical coupling reaction exists between glucosamines in the hyphal walls of the fungus with phenolic

TABLE 5.7.   Faunal C, Microbial Biomass, Mineralized C and N, and N and P under Stubble Mulch and No-Till Treatments of the Fallow Phase of Dryland Wheat Plots[a]

Variable	Treatment	Date				
		8 June	6 July	2 August	23 August	13 September
Faunal carbon[b]						
Collembola	Stubble mulch	6.69	1.03	2.18	0.61	1.49
× 100	No till	5.58	8.29	4.01	1.60	1.49
Acari	Stubble mulch	3.77	0.47	0.91	0.40	0.92
× 10	No till	3.60	3.31	1.24	1.32	0.54
Holophagus						
nematodes	Stubble mulch	0.88	0.46	1.12	0.51	0.56
× 10	No till	0.48	1.69	0.96	0.57	0.32
Protozoa	Stubble mulch	1.76	0.67	0.74	1.96	1.92
× 1	No till	2.20	0.96	0.93	2.29	1.92
Microbial biomass[c]						
Carbon	Stubble mulch	245	204	271	186	255
	No till	329	273	299	194	256
Nitrogen	Stubble mulch	81	57	53	60	56
	No till	92	72	49	62	48
Phosphorus	Stubble mulch	5.7	7.5	10.1	4.6	7.2
	No till	5.5	6.8	10.1	4.9	7.1
Mineralized C and N[d]						
Respired C	Stubble mulch	52	77	57	42	68
(0–10 days)	No till	97	110	98	41	67
Respired C	Stubble mulch	49	42	56	53	24
(0–20 days)	No till	80	69	84	57	24
Mineralizable N	Stubble mulch	10.63	6.39	5.79	2.62	1.95
(0–20 days)	No till	12.68	5.79	8.90	−3.3	−1.38
N and P[e]						
NH_4^+ - N	Stubble mulch	3.1	1.0	2.0	1.8	0.8
	No till	4.0	1.8	4.6	1.2	1.4
NO_3^- - N	Stubble mulch	7.2	10.6	16.3	17.8	16.1
	No till	7.8	23.8	21.6	45.1	46.2

[a]From Elliott *et al.* (1984). Reprinted by permission of Kluwer Academic Publishers.
[b]Soil fauna biomass C (kg C ha^{-1} to 10 cm) for four categories. (Note differences in the multiplier for each category.)
[c]Microbial biomass C, N, and P (kg element ha^{-1} to 10 cm).
[d]Mineralizable C (CO_2 - C) and NO_3-N as Kg ha^{-1} to 10 cm in unchloroformed 0 to 10- or 10- to 20-day incubations of soils sampled from the field.
[e]NH_4-N, and extractable inorganic and organic P amounts (kg ha^{-1}) in the top 10 cm of soil.

compounds released during lignin degradation from leaf and root tissues. An additional possibility, little investigated yet, is the apparently widespread occurrence of interspecific physical linkages, enabling transfers of nutrients via mycorrhizae of various annual and perennial plants (Chiariello *et al.*, 1982; Read *et al.*, 1985; Read, 1991). Physical, chemical, and biological contacts may be operating simultaneously in mycorrhizal-mediated interactions.

Some of the latter examples may seem a bit removed from the general theme: the role of soil fauna in soil processes. However, it is apparent from studies by Warnock *et al.* (1982), Moore *et al.* (1985), and Curl and Truelove (1986) that soil mesofauna, e.g., Collembola, show considerable preference for, and have an impact on, AM fungal growth, just as they do for saprophytic fungi (Newell, 1984a,b). This undoubtedly extends to nematodes (Ingham *et al.*, 1985) and soil amoebae as well (Chakraborty and Warcup, 1983; Chakraborty *et al.*, 1983; Gupta and Germida, 1988).

SUMMARY

The major lesson to be learned as a soil ecologist is one of paying attention to details, yet considering them in a holistic perspective. Certainly we are past the time when measurements of the "soil biomass" (referring to the microbial biomass) alone, by whatever method, are considered adequate (Coleman, 1994a). Small groups of organisms, perhaps highly aggregated within the ecosystem, may be facilitating (or retarding) the turnover of other organisms or major nutrients, such as nitrogen, phosphorus, and sulfur.

Decomposition rates, along with nutrient dynamics, soil respiration, and formation of soil structure, are integrating variables. They are generalized measurements of the functional properties of ecosystems, and they summarize the combined actions of soil microflora, fauna, abiotic variables and resource quality factors. Litter breakdown rates can be compared, using simple first-order models so that rate variations between ecosystems or between different substrates may be compared. Litter breakdown rates are easily measured using bagged leaf litter ("litterbags"). Decomposition per se is due to microbial activities, but experiments show that fauna have a strong influence on litter breakdown rates, especially for more resistant substrates. The interaction between microflora and fauna is especially important for nutrient cycling mechanisms. Organic matter dynamics are strongly influenced by soil fauna. Termites and earthworms are well-known for their influences on nutrient dynamics, soil organic matter, and soil structure. But the entire soil fauna is involved in these processes and, through their interactions with soil microbes, must be considered in a holistic perspective.

6 | *Detritivory and Microbivory in Soils*

INTRODUCTION

The traditional studies of food webs and food chains began with pioneering efforts of Summerhayes and Elton (1923) in Spitsbergen. This early study explicitly linked detrital biotic interactions with other parts of the terrestrial and aquatic food web (Fig. 6.1). Work on detrital food webs progressed slowly for the next 20 years, although Bornebusch (1930) carried out some pioneering studies of detrital food webs and their energetics. Further insights were gained from the studies of Lindeman (1942), who developed the concept of trophic levels.

By the early 1960s, more holistic study of energetics and interactions of organisms in ecosystems was clearly needed. This led to the ambitious effort known as the International Biological Program (IBP). The overall intent was to bring working groups together, addressing how carbon and energy flow in a wide range of terrestrial and aquatic ecosystems, with the ultimate goal being a better understanding of how ecosystems work and how they could be manipulated for the benefit of mankind. The IBP found that for a wide range of grassland, desert, and forested ecosystems, the net flow into the aboveground grazing (consumer) component is only 5% or less, with the remainder

a

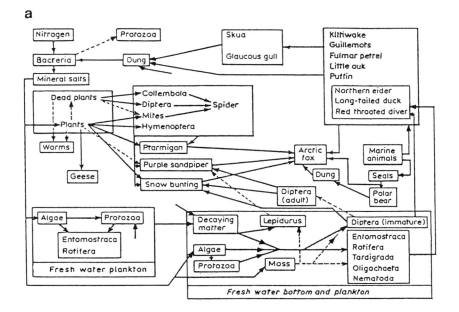

b

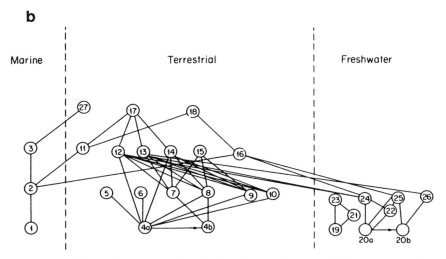

FIGURE 6.1 Arctic food web (a) described by Summerhayes and Elton (1923) and as diagrammed (b) by Pimm (1982). Legend (1) plankton, (2) marine animals, (3) seals, (4a) plants, (4b) dead plants, (5) worms, (6) geese, (7) Collembola, (8) Diptera, (9) mites, (10) Hymenoptera, (11) seabirds, (12) snow bunting, (13) purple sandpiper, (14) ptarmigan, (15) spiders, (16) ducks and divers, (17) arctic fox, (18) skua and Glaucous gull, (19) planktonic algae, (20a) benthic algae, (20b) decaying matter, (21) protozoa, (22) protozoa, (23) invertebrates, (24) Diptera, (25) other invertebrates, (26) *Lepidurus*, and (27) polar bear. From Pimm (1982) and Pimm and Lawton (1980).

entering the detrital-decomposer food web (Coleman *et al.*, 1976). This research led to several post-IBP studies in North America and Europe to follow up on the initial results.

In the late 1970s and 1980s, a series of investigations of detrital food webs were carried out in the semiarid and arid grasslands and desert lands of Colorado and New Mexico (Coleman *et al.*, 1977, 1983; Parker *et al.*, 1984; Whitford *et al.*, 1983; Hunt *et al.*, 1987; Moore *et al.*, 1988) (Fig. 6.2). These studies and several in Sweden (Persson, 1980; Bååth *et al.*, 1981) and in the United Kingdom. Anderson *et al.*, (1985) found that microbial/faunal interactions have significant impacts on the nutrient cycles of the major nutrients; nitrogen, phosphorus, and sulfur (Gupta and Germida, 1989). Some of these studies used assemblages of a few species in microcosms, but were beginning to delineate the mechanisms that are important in soil systems in general. Among the fauna, the protozoa were often overlooked, despite the findings by Cutler *et al.* (1923) that there are important predator–prey interactions between protozoa and bacteria in soils. Clarholm (1985) noted that soil protozoa are avid microbivores and turn over an average of 10–12 times in a growing season, in contrast to many other members of the soil biota, which may turn over only once or twice in a ca. 120- to 140-day growing or activity season. These findings were further extended (Kuikman *et al.*, 1990), with the observation that nitrogen uptake by plants may increase from 9 to 17% when large inocula of protozoa are present. The demographics and microbial/faunal interactions provide much of the driving force in the models of nitrogen turnover in semiarid grasslands (Hunt *et al.*, 1987) and arable lands (Moore and de Ruiter, 1991).

More recent studies have noted the more complex nature of food webs when detrital components are included (Polis, 1991; Hall and Raffaelli, 1993). DeAngelis (1992) in a treatise on nutrient cycling, devoted an entire chapter to the nutrient interactions of detritus and decomposers. His ideas have provided several insights into decomposition/nutrient cycling processes. This chapter addresses several aspects of soil biota and nutrient cycling in soils, namely demography and "hot spots of activity" which are often overlooked in energetics studies of soil systems. These factors are crucial to understanding how organisms and soils interact and how they contribute to ecosystem function.

PHYSIOLOGICAL ECOLOGY OF SOIL ORGANISMS

Given the physiological ecology of the microbes and fauna involved, are long food chains possible? There are several theoretical reasons

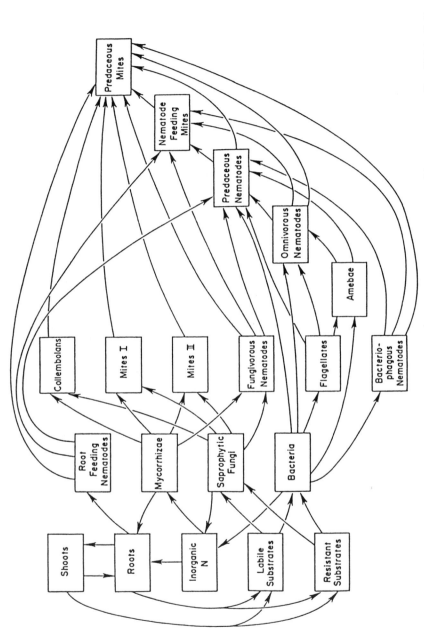

FIGURE 6.2 Representation of detrital food web in shortgrass prairie. Fungal-feeding mites are separated into two groups (I and II) to distinguish the slow-growing cryptostigmatids from faster-growing taxa. Flows omitted from the figure for the sake of clarity include transfers from every organism to the substrate pools (death) and transfers from every animal to the substrate pools (defecation) and to inorganic N (ammonification) (from Hunt *et al.*, 1987. The detrital food web in a shortgrass prairie. *Biol. Fertil. Soils* **3**, 57–68. Copyright © 1987 Springer-Verlag. Reprinted with permission.).

why long food chains could be expected. Let us take, as an example, the energetically most dominant interactions between microbes and fauna that occur in many terrestrial ecosystems, summarized by Hunt *et al.* (1987, Fig. 6.2). The flow of organic carbon or nitrogen moves from initial organic substrates (labile or resistant) to the primary decomposer, either bacteria or fungi, then on into microbivorous microfauna (flagellates and amebae) or microbivorous mesofauna (feeding on fungi), then to omnivorous or predaceous nematodes, and on to nematode feeding mites and predaceous mites. Further predation on the mites by ants (E. O. Wilson, personal communication, 1993.) or lithobiomorph Chilopods (centipedes) is possible, although not explicitly represented by Hunt *et al.* (1987). There are at least eight links in the bacterial-based detrital food chain, with considerable evidence of omnivory. For example, many fungivorous mites require a nematode "supplement" to complete their life cycles (Walter *et al.*, 1991). Note that this is a rather ecosystem-specific diagram. One could draw another for decomposition in a coniferous or oak/beech forest, with a significant proportion of the total decomposition being mediated by ectotrophic mycorrhizae, operating perhaps in competition with the saprophytic fungi (Gadgil and Gadgil, 1975).

For desert and estuarine food webs, reviewed by Hall and Raffaelli (1993), the detrital food chain length noted earlier is comparable to the average length of five to seven links (Polis, 1991), with eight maxima recorded. In contrast, Hairston and Hairston (1993) assert that the usual food chain length in detrital systems seldom exceeds three. As noted in the following discussion, these long chain lengths of five to seven links are not only feasible, but are thermodynamically possible at several times and in several locations in the soil matrix, particularly the rhizosphere, and other "hot spots" of activity. What levels of taxonomic resolution are both most useful and appropriate for detrital food web studies? Our inability to sort out the details of microbial taxonomy *in situ* and our limited knowledge of many of the soil invertebrates, particularly the immatures (Behan-Pelletier and Bissett, 1993) require the use of rather coarse functional groups for taxonomy of the soil biota. Interestingly, this sort of separation enabled Wardle and Yeates (1993) to identify competition and predation forces operative in an assemblage of detritus/microbial/nematode trophic groups in an agricultural field. Using a correlation analysis, they noted that predatory nematodes reflected most closely the changes in primary production and that the microbivorous nematodes seemed to be more dependent on substrate quality in the microbial (bacterial and fungal) community.

ENERGY AVAILABLE FOR DETRITAL FOOD CHAINS AND WEBS

The energy available for detrital food chains is considered next. If one considers the variance or range around the mean values of the assimilation and production efficiencies of the biota (Table 6.1), the amount of energy can be calculated which will move from primary decomposers all the way up the food chain. There is indeed energy to spare for such elaborate food chains. Using the maximal values for production efficiency, such as 70% for bacteria (Payne, 1970) and 80% for soil amoebae and flagellates (Humphreys, 1979), and moving on to protozoan-consuming nematodes (Fig. 6.3), which doubtless occur in certain hot spots, e.g., at the zone of elongation of a growing root, then there is an adequate amount of carbon available for passage through the four- and five-membered detrital food chains of interest. Considerable omnivory is prevalent in these soil systems (DeAngelis,

TABLE 6.1. Physiological Data on Major Biotic Groups in Soil[a]

Tropic group	Fraction of food assimilated			Production: assimilation ratio		
	max	$\overline{X}$	min	max	$\overline{X}$	min
Bacteria	?	1.0	<0.01?	0.7	0.4	<0.01?
Saprophytic fungi		1.0		0.7	0.4	
Arbuscular mycorrhiza		?		0.8?	0.4	
Amoebae		0.95		0.8	0.4	
Flagellates		0.95		0.8	0.4	
Phytophagous nematodes		0.25		0.5?	0.37	
Fungivorous nematodes		0.38		0.5?	0.37	
Bacterivorous nematodes		0.6		0.5?	0.37	
Omnivorous/predaceous nematodes		0.55		0.5?	0.37	
Fungivorous mites [b]		0.5		0.5?	0.35	
Fungivorous mites [c]		0.5		0.5?	0.35	
Nematophagous mites		0.9		0.5?	0.35	
Predaceous mites		0.6		0.5?	0.35	
Collembola		0.5			0.35	
Enchytraeids		0.28			0.4	
Earthworms		0.2			0.45	
Termites		0.4?			0.15?	

[a]Modified from Humphreys (1979), Hunt et al. (1987), Payne (1970), and de Ruiter et al. (1993).

[b]Rapid growth strategy.

[c]Slow growth strategy.

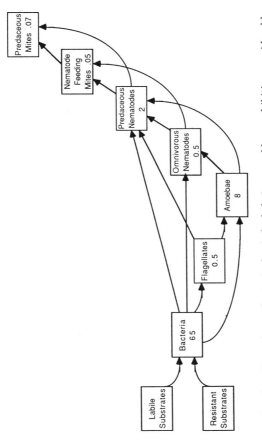

FIGURE 6.3 Calculations of annual carbon flows along a bacteriophagic food chain assemblage, exhibiting considerable omnivory. Average standing crops are indicated. Flows via protozoan feeding are estimated as probably two to three times greater than via nematodes in that ecosystem. Proportions may be reversed in lower pH forested systems, and more flows via fungi (not shown). From Hunt *et al.*, 1987. The detrital food web in a shortgrass prairie. *Biol. Fertil. Soils* **3**, 57–68. Copyright © 1987 Springer-Verlag. Reprinted with permission.

1992). The protozoa and nematode feeding pathway highlighted in Fig. 6.3 (Hunt *et al.*, 1987) accounted for 37% of the total nitrogen mineralization and some 82% of the total mineralization due to soil fauna. Similar percentages were obtained for a wide range of agroecosystems in the United States and Europe (De Ruiter *et al.*, 1993, 1995).

The relative contributions of the soil fauna to microbial turnover and nutrient mineralization are directly related to the demographies of the soil biota (Coleman *et al.*, 1983; 1993), as noted for average standing crops and energetic parameters and turnover times per year for microorganisms, micro-, meso-, and macrofauna in a grassland, and a no-tillage agroecosystem (Coleman *et al.*, 1993) (Table 6.2). Thus the protozoa, and naked amoebae in particular, turn over 10 or more times per season and consume several times their mass of living microbial tissues. The microbes and several other faunal groups have much lower turnover rates on average. Although the amoebae are considered to be primarily bacterial feeders, there are important instances when other amoebal species will feed on protoplasm in fungal hyphae or even on the fungal spores themselves (Chakraborty and Warcup, 1983; Chakraborty *et al.*, 1983). When considered in combination with the information in Table 6.1 on the range of assimilation and production efficiencies, the impacts of these small organisms are very marked. It should be noted that extensive studies in Sweden on arable lands (Andrén *et al.*, 1990) have reached similar conclusions.

ARENAS OF INTEREST

Soils can be considered most profitably as the extremely heterogeneous entities they are. This requires that we "let the soil work for us" (Elliott and Coleman, 1988) and stratify, in a statistical sense, the regions of the soil that are "hot spots" of activity. These zones include the rhizosphere, aggregates, litter and organic detritus, and the "drilosphere," which is that portion of the soil volume influenced by secretions of earthworms (Bouché, 1975) (Fig. 6.4). Each region is a relatively small subset of the total soil volume, but may contain a preponderance of numbers and more importantly, activity of the soil biota (Beare *et al.*, 1995). Examples include the 5–7% of the total soil that was root influenced or rhizosphere, in extensive pot trials of Ingham *et al.* (1985), contained a majority (greater than 70%) of the bacterial and fungal-feeding nematodes. Ingham *et al.* (1985) also measured higher biomasses of rhizosphere bacteria in microcosms with large numbers of microbivorous nematodes (greater than 4000 per g of rhizosphere soil) than in microcosms without these nematodes. Yet the

TABLE 6.2. Average Standing Crop and Energetic Parameters for Microorganisms, Mesofauna, and Earthworms in a Lucerne Ley and Georgia No-Tillage Agroecosystem[a]

	Naked amoebae	Flagellates	Ciliates	Bacteria	Fungi	Microbivorous nematodes	Collembola	Mites	Enchytraeids	Earthworms
Typical size in soil	30 μm	10 μm	80 μm	0.5–1x 1–2 μm	Ø 2.5 μm 1.0–5.5 μm	Ø ~ 40 μm	Ø 5000 μm	Ø 1000 μm	Ø 1000 μm	Ø 5000 μm
Mode of living	In water films on surfaces	Free-swimming in water films		On surfaces	Free and on surfaces	In water films, free, and on surfaces	Free	Free	Free	Free in soil
Biomass (kg dw ha^{-1})	95%	5% 50[b]	<1%	500–750[c]	700–2700[d]	1.5–4[e]	0.2–0.5[e]	2–8[e]	1–8[e]	25–50[e]
% active	0–100			15–30	2–10	0–100	80–100	80–100	?	0–100
Estimated turnover times, season^{-1}		10		2–3	0.75	2–4	2–3	2–3	?	3
No. of bacteria division^{-1} x 10^{-3}	3–8	0.6–1	20–2000							
Minimum generation time in soil (hours)		2–4		0.5	4–8	120	720	720	170	720

[a]Modified from Clarholm (1985), Hendrix *et al.* (1987), and Beare *et al.* (1992). Reprinted with permission from Coleman *et al.* (1993). Copyright Lewis Publishers, an imprint of CRC Press, Boca Raton, Florida.

[b]MPN technique.
[c]Direct counts plus size class estimations.
[d]Direct estimation of total hyphal length and diameter.
[e]Extractions and sorting.

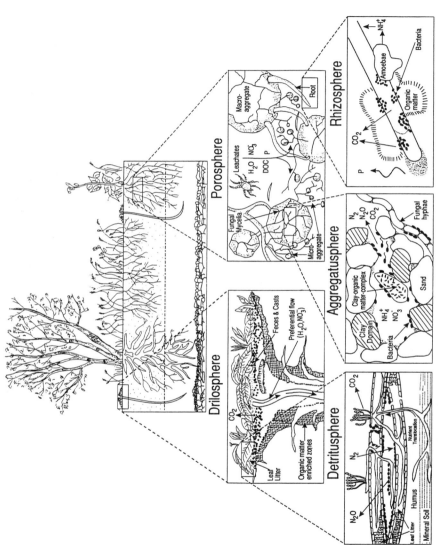

FIGURE 6.4 Arenas of activity in soil systems. These "hot spots" of activity may be <10% of the total soil volume, but represent >90% of the total biological activity in most soils worldwide (from Beare *et al.*, 1995). Reprinted by permission of Kluwer Academic Publishers.

extent of mineralization of nitrogen in the microcosms with nematodes reflected that they were ingesting large quantities of microbes as well. Thus there was a net enhancement of microbial production in a fashion similar to that measured by Porter (1975), who found a net stimulation of phytoplankton growth after the cells had undergone transit through the guts of *Daphnia* sp. in freshwater incubations. As an example of the dynamic nature of shifting hot spots, Griffiths and Caul (1993) found that more nematodes were active in the rhizosphere and that they moved readily to new concentrations of fresh organic matter (leaf litter) in short-term trials. Other examples of hot spots which have shown enhanced microbial activity include the drilosphere and worm castings, which show enhanced carbon and nitrogen (Syers *et al.*, 1979a; Daniel and Anderson, 1992) and phosphorus mineralization (Syers *et al.*, 1979b; Lavelle *et al.*, 1992). Another center of activity is the aggregatusphere (Fig. 6.4), or region of micro- and macroaggregates (Elliott, 1986; Elliott and Coleman, 1988; Beare *et al.*, 1995). This zone of influence is less well studied energetically, yet is a major source of some of the dynamical, yet highly patchy, behavior found in soils. Foster (1985) and Foster and Dormaar (1991) have demonstrated, using electron microscopy (Fig. 3.2), amoebal pseudopodia extending into very small porenecks and pores (only a few tenths of micrometers in diameter) in well-structured soil, attacking bacterial colonies which seemed to be inaccessible to the smallest nematodes and amoebae or other protozoa. A combined approach to rhizosphere and soil cracks for locations of "hot spots" of labile organic matter was used by van Noordwijk *et al.* (1993) to good effect.

A HIERARCHICAL APPROACH TO ORGANISMS IN SOILS

Because of the need to deal with soil heterogeneity in space and time, arenas of interest, noted earlier, are represented in Fig. 6.4 (Beare *et al.*, 1995) showing the volumes and biotic groups of concern. The aggregatusphere shows bacteria, amoebae, and some nematodes having varying degrees of success in gaining access to the prey biota of interest (Vargas and Hattori, 1986). Moving up to a coarser level of resolution, to the rhizosphere, which is a few millimeters or less in scale, one is able to see the microbes and fauna associated with them and the considerable feeding and activity which has been documented numerous times. The activities are strongly influenced by abiotic, i.e., wetting and drying events, and the intrusion of new organic substances from growing root tips (Cheng *et al.*, 1993) or from deposited feces from microarthropods, enchytraeids, or other mesofauna. The next level of resolution expands from many centimeters to several meters across the landscape, when any of the macrofauna, such as earthworms or burrowing beetles, come into play. There is then a qualitative shift, brought about by the ingestion of soil, which includes considerable amounts of micro- and mesobiota, e.g., protozoa and nematodes (Yeates, 1981; Piearce and Phillips, 1980) as food. Interestingly, even with earthworms, the drilosphere, *sensu stricto* is only 2–3 mm in thickness (Bouché, 1975) but the burrow extends laterally for many centimeters or meters through the soil. As a consequence of this activity, there can be major short-term decreases in the viability of the existing biota, but possibly longer term stimulation by enhanced microbial activity, as noted earlier, and also from the considerable input of mucopolysaccharide-containing mucus (Marinissen and Dexter, 1990). In addition, considerable amounts of ammonia and urea, as nitrogenous end products of metabolism, may be voided either externally, through nephridiopores, or internally into the gut cavities of earthworm genera which have that mode of nitrogen excretion (Lavelle *et al.*, 1992). In tropical regions, certain endogeic earthworms will process and assimilate end products of the breakdown (from 2 to 9%) of soil organic matter in a wide range of ecosystems (Lavelle and Martin, 1992). Similar sorts of activities may be catalyzed by certain termites, particularly those in the advanced family Termitidae, which are truly geophagous, and will utilize considerable amounts of soil organic matter, deriving significant amounts of nutrition from this low-quality substrate by processing the organic matter in a high pH chemical milieu in the region between the midgut and the first proctodaeal segment (Bignell, 1984). The additional influence of microbial enzymes on insect digestive processes and indeed enhancement of nitrogen fixation in downed branches and logs (Martin, 1984) are well known. Finally, the impacts of ant and termite nests are significant and certainly have an influence at the landscape scale.

FUTURE RESEARCH PROSPECTS

It is becoming more and more imperative to bring small working groups, or teams of investigators, together to make further progress in food web studies. The real breakthroughs are certain to come from efforts which include the more transitional fauna between above- and belowground, such as ants, dipteran larvae and ground beetles, or cryptozoans, such as the myriapoda, linking them to the truly belowground fauna and microbes.

Various techniques noted in earlier chapters should be extended as well. Stable isotopes, introduced in an initially enriched substrate, such as labeled glucose or acetate, will be useful in delineating food webs. The increasing miniaturization of sensors, so that one can carry out microcalorimetry (Battley, 1987) at localized microsites, will enable us to measure direct energetic transformations more readily *in situ*.

Hall and Raffaelli (1993) suggest two major areas of food web research which could be most profitably followed: (1) focusing on community assembly, and (2) documenting the strength of trophic interactions between elements in webs. We suggest that a useful approach will include a melding of the two objectives in terms of documenting the extent of soil food webs, and the relative impacts of the trophic interactions at the various hierarchical levels of organization and location in the landscape (Coleman and Schoute, 1993). For example, does the soil system in the absence of earthworm or termite activity operate at more or less of a background or maintenance level? When the macrofauna move through the soil matrix, literally consuming and chaotically reassembling it, does this represent a more intensive level of activity? All this awaits further investigation, in experimental fashion, by the new generation of soil ecologists.

SUMMARY

An interesting convergence occurs in aboveground and belowground portions of detrital food webs. In both locations, particularly in arid habitats (i.e., in deserts), the food webs are long (seven to eight membered) and show extensive amounts of omnivory.

By including members of the microfauna (protozoa) and mesofauna (microbivorous nematodes) which have been overlooked often in the past, ample amounts of food, as secondary production, pass up the food chains, as production efficiencies may reach or exceed 70% in various "hot spots" such as rhizospheres, drilospheres, or at any other concentrations of reduced, labile organic matter.

Future Developments in Soil Ecology

INTRODUCTION

There are several areas of interest to ecologists in general and to soil ecologists in particular as we enter the 21st century. The effects of soil processes and soil biota on global change, particularly with relation to global greenhouse gases, is of concern to resource managers and globally oriented ecologists (Coleman *et al.*, 1992). More recently, there has been a rising current of interest in soils and biodiversity. Within terrestrial ecosystems, soils may contain some of the last great "unknowns" of many of the biota. This includes such relatively well-studied fauna as ants (Hölldobler and Wilson, 1990). A final area concerns the roles of soils and the "Gaia" mechanism, or overall global ecosystem functional concept.

ROLES OF SOILS IN GLOBAL CHANGE

Soils are probably the last great frontier in the quest for knowledge about the major sources and sinks of carbon in the biosphere. The direct effects of deforestation on global patterns of carbon cycles are relatively minor; the effects of changed sink strengths, with defor-estation decreasing the rates of CO_2 uptake, may be much larger. Another source of carbon input to the atmosphere has come from the oxidation of soil organic matter during cultivation of native lands, such as the Great Plains region of North America and the Eurasian steppes of eastern Russia (Wilson, 1978; Houghton *et al.*, 1983). The standing stocks of soil carbon are twice as large as all of the standing crop biomass of all of the terrestrial biomes combined (Fig. 7.1) (Anderson, 1992). However, the plant/soil systems are strongly coupled, and the rates of inflows and outflows are significantly con-trolled by rates of above- and belowground herbivory in forests (Pastor and Post, 1988) and in grasslands (Schimel, 1993). The feedback effects of the principal greenhouse gases, namely carbon dioxide, methane, and nitrous oxide, are very large (Mosier *et al.*, 1991; Rogers and Whitman, 1991), with the effects of carbon dioxide being some 56% of the total impact (Anderson, 1992). However, the rate of increase of methane is almost twice that of CO_2 (Houghton *et al.*, 1987; Houghton, 1990) so it is being closely observed by atmospheric

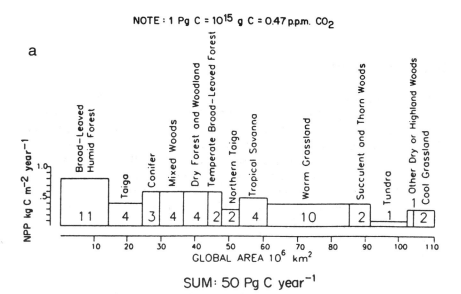

FIGURE 7.1 Pools and fluxes of carbon in major terrestrial ecosystem types. (a) Distribution of net primary production; (b) biomass, and (c) soil carbon pools. The total area occupied by each ecosystem type is represented by the horizontal axis with flux or density of the vertical axis; the area is therefore proportional to the global production or storage in each ecosystem type. (Source: Lashof (1989) based on data in Olson *et al.* (1983) and Zinke *et al.* (1984) (from Anderson, 1992).

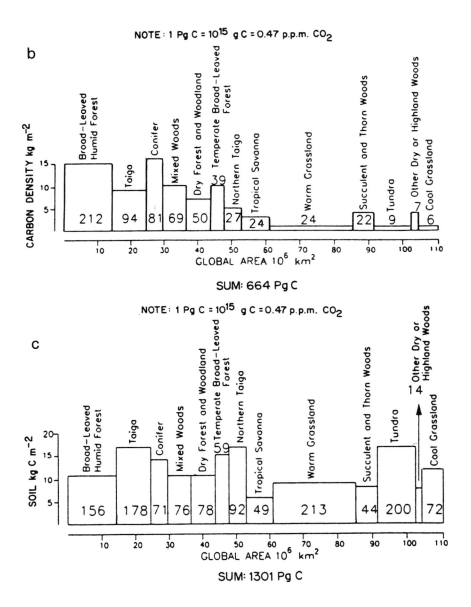

b

NOTE : 1 Pg C = 10^{15} g C = 0.47 p.p.m. CO_2

SUM: 664 Pg C

c

NOTE: 1 Pg C = 10^{15} g C = 0.47 p.p.m. CO_2

SUM: 1301 Pg C

scientists. One of the major concerns of scientists interested in global change is the extent of involvement by soils and soil processes in the evolution of greenhouse gases and the roles of soil biota and organic matter in the global carbon cycle (Lal *et al.*, 1995).

FUNCTIONAL DYNAMICS OF SOILS IN RELATION TO GLOBAL CHANGE

We examine next the ways in which soils operate over ecological and geological time spans and how they may be influenced by, or have an effect on, global change processes. Soil development and change may be viewed as the result of the basic processes of additions, removals, transformations, and translocations (Anderson, 1988). A given landscape will experience runon, runoff, transformations, transfers up and down in the profile, and additions and losses either aerially or pedologically (Fig. 7.2). These processes may be very dynamic for processes such as movement of soluble salts, which vary within seasons, or be measured in thousands of years, e.g., for clay

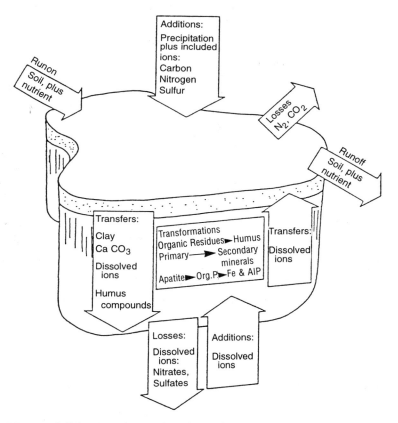

FIGURE 7.2 Soil-forming factors based on the concepts of Simonson (1959) as described by Anderson (1988) (from Stewart *et al.*, 1990).

weathering processes. The microbial portion of the organic matter cycle will have mean net turnover times of 1–1.5 years, whereas humification processes, such as the interactions of clay/humic compounds, may be considered intermediate (centuries) in the time scale (Stewart *et al.*, 1990) (Table 7.1). These processes can be envisioned readily via the carbon, nitrogen, and phosphorus submodels of the Century model (Fig. 7.3). This model was developed to simulate the additions and losses in agricultural lands and grasslands worldwide (Parton *et al.*, 1987, 1989a), but has now been extended to a wide range of ecosystems, including tundra and taiga (Smith *et al.*, 1992) and tropical ones as well (Parton *et al.*, 1989b, Schimel *et al.*, 1994).

ROLES OF SOILS IN THE GLOBAL CARBON CYCLE

What patterns and processes of global change are most likely to affect the global carbon cycle in soils? What are the effects of climate change on vegetation; are there possible changes in sink strengths (pools of organic matter, active roots, etc.) in various parts of the globe? Do we know enough about the dynamics of carbon in the 13 or more major biomes which comprise the terrestrial biosphere? For example, consider the size of the live biomass in a broad-leaved humid forest, which amounts to 212 petagrams (pg $=10^{15}$ g) versus warm grasslands which have only 24 pg live biomass. When comparing the amounts of soil carbon, stored vs carbon in live biomass, there is relatively less storage in the humid broad-leaved forest (156 pg), giving a ratio of 212/156 or 1.36 more in the live biomass. Warm grasslands, with 213 pg in soil organic matter, have a ratio of 24/213 or 0.11 in biomass versus that in the soil organic matter. Tundra, with only 9 pg in

TABLE 7.1.   Grouping of Soil-Related Processes and Components Based on Time[a]

Highly dynamic	Dynamic	More static, slow
Soluble nutrients	Adsorbed nutrients	Nutrient reserves in minerals
Active or soluble organic matter	Labile organic matter adsorbed to clay	Chemically stabilized organic matter
Solution and movement of soluble components	Weathering of carbonate minerals	Weathering of silicates and clay minerals
Microbial growth	Microfauna and mesofauna Plant growth	Vegetation, i.e., forest

[a]From Stewart *et al.* (1990).

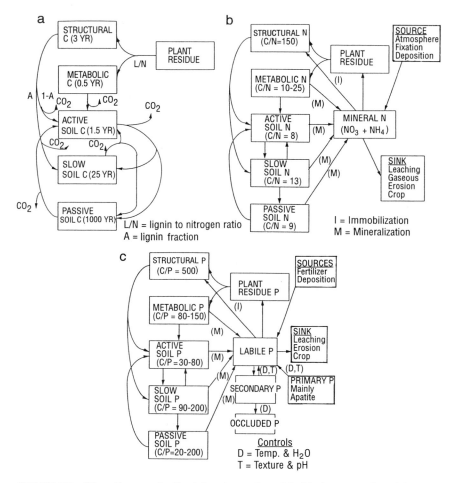

FIGURE 7.3   Flow diagram for the (a) carbon submodel, (b) nitrogen submodel, and (c) phosphorus submodel of CENTURY (adapted from Parton *et al.*, 1987, 1989b) (from Stewart *et al.*, 1990).

live biomass vs 200 in the soil organic matter, has a ratio of only 0.05 in living biomass vs soil organic matter. What are the climatological vs plant physiological and microbiological implications of such differences in these widely different biomes? Research in this area requires considerable effort in soil science in addition to microbial ecology because we are faced with problems of measuring substrate quality, covered earlier, and its feedback effects on future primary production and nutrient dynamics. Of course the modes of growth of grasses versus trees are also influential because more of the total growth effort is invested below ground in both grassland and tundra soils.

Soil science meetings and congresses have featured discussions about the roles of soils in the global carbon cycle. For example, the International Soil Science Society Congress in 1994 had a major symposium on the topic "Impact of soil on the carbon cycle in managed and natural ecosystems." Several papers addressed key aspects of the terrestrial carbon cycle: carbon fixation by primary production and then mechanisms for sequestering the carbon during organic matter decomposition and transformation processes or mechanisms for mineralization via human-induced or natural processes. The following is a brief synopsis of the major points developed in several papers presented at the symposium. The first part of the symposium was concerned with the results of a workshop on "Management of carbon in tropical soils under global change: Science, practice and policy" which was held in Nairobi, Kenya (Kimble and Levine, 1994). Although focused on tropical countries, the workshop covered many aspects of global carbon problems, with its central concern being that soil organic carbon is the second largest pool in the terrestrial organic carbon cycle, with about 1550 pg involved.

Concerns about imbalances in the global carbon cycle are not new; rapidly increasing amounts of carbon dioxide entering the atmosphere from human activities, including burning of fossil fuels, were noted first 1 century ago (Arrhenius, 1896). Since then, interest in the rates of the flow of carbon and amounts sequestered in various pools in the biosphere has waxed and waned. For example, Plass (1956) expressed concern about the amounts of carbon dioxide being released by the burning of fossil fuels worldwide. The relative amounts of soil organic matter being "mined out" by extensive cultivation throughout the major "breadbaskets" of the world, such as in the North American Great Plains, the former Soviet Union, and Canada, are quite large, perhaps up to 40% of the surface layers (Haas *et al.*, 1957; Coleman *et al.*, 1984), More recently, Mann (1986) concluded in a survey of 625 soils studied pairwise, cultivated vs noncultivated on the same soil type, that 20% or more of carbon was lost over decadal time spans from soils with high amounts of carbon (ranging from 6 to 16 kg/m^2). Interestingly, she noted that modest gains occur in soils that are initially very low in soil organic carbon, such as very sandy textured ones, if they are put into cultivation. A significant amount of carbon fixation, and subsequent movement into the soil organic matter, in the surface to 30 cm depth, will occur over a several years time span. If extensive application of fertilizers is required to achieve these gains, then the global carbon balance overall is still toward the positive side in terms of carbon costs for fossil fuel-derived nitrogen, for example.

The CO_2 balance of soils has been modeled by various groups. In studies on oxisols (ferralsols in the FAO classification) in Surinam in South America, van Breemen and colleagues found that geographic data on land use and changes in land use were necessary to make detailed predictions of the fluxes of carbon into and out of specific soil types (Van Breemen, 1992). However, they found that process-oriented models were preferable to the more static bookeeping types of models to estimate carbon fluxes into and out of soils.

PROBLEMS IN MODELING SOIL CARBON DYNAMICS

A more general problem yet faces us as soil ecologists. One of our current needs is to "model the measurable" rather than "measure the modellable" (Elliott, 1994). There are pools in models such as the Century, mentioned earlier, which are more easily conceptualized than actually measured. A more readily measurable entity is the labile pool, consisting primarily of the microbial biomass. The intermediate and longterm pools, existing from decades to millennia, are very difficult to measure directly, and much new work is under way to more effectively isolate and characterize these pools by a variety of methods (Beare *et al.*, 1994a,b; Cambardella and Elliott, 1994). This problem requires integration across several levels of resolution, dealing with numerous human activities in sociology and economics which have a direct impact on soil management. These include the concept of the effectiveness of management of carbon resources, which is inversely related to the cost of subsidizing the lost functions of organic matter (Fig. 7.4) (Woomer and Swift, 1994). The effectiveness of carbon resource management decreases with the sequential loss of constituents and the subsequent loss of function as land use intensifies without subsidizing lost organic matter. Elliott and colleagues (1994) urged soil ecologists to isolate functional soil organic matter fractions and to determine their roles in soil processes in order to understand the mechanisms controlling soil processes. This includes the mechanisms and processes involved in the formation and turnover of macro- and microaggregates in a wide range of soil types worldwide (see, e.g., Beare *et al.*, 1994a,b). Indeed, chemical, microbiological, and macrobiological characterization of physically isolated fractions may provide the best opportunity for identifying functional pools of soil organic matter. For example, each major category of soil biota has a significant effect on one or more aspects of soil structure, including the production of organic compounds that bind aggregates and the hyphal entanglement of soil particles (microflora), producing fecal pellets and creating biopores (meso- and macrofauna) (Hendrix *et al.*, 1990;

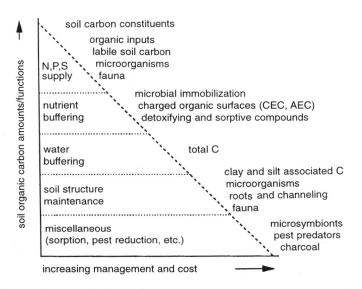

FIGURE 7.4   The functional role of soil organic matter within an ecosystem depends on the intensity with which that system is managed (personal communication from E. T. Elliott; modified from Woomer and Swift, 1994).

Linden *et al.*, 1994). A complete list of influences of soil biota is given in Table 4.8 (Hendrix *et al.*, 1990).

BIOLOGICAL INTERACTIONS IN SOILS, AND GLOBAL CHANGE

Perhaps the principal element of the global change scenario is the steadily increasing annual temperature, which rises about 0.1° C annually. As this increase occurs, there should be a perceptible increase in the loss of soil carbon (Schimel *et al.*, 1994; Bouwman, 1990; Scharpenseel *et al.*, 1990; Jenkinson *et al.*, 1991; Lal *et al.*, 1995). A countervailing tendency will exist with effects of CO_2 fertilization, enhancing plant primary production. However, several authors have noted the further constraints of other limiting resources to growth, such as mineral nutrients (Pastor and Post, 1988; Schimel, 1993). More holistic modeling efforts of changes in soil organic matter, particularly ones which include soils, plants, herbivores, and detritivores together, are more realistic in their outcomes than those which model the plants, heterotrophs, and soil carbon pools separately (Schimel, 1993).

A Hierarchical View of Soil Ecology

Hierarchical approaches to ecological research have been increasing in interest since the 1980s (Allen and Starr, 1982; O'Neill et al., 1986). There are a number of levels of resolution organized in a landscape context, utilizing ecosystem-level criteria, such as movement of water and materials within a watershed (Fig. 7.5). The conceptual diagram ranges from an entire watershed at the widest scale down to microsites on roots and leaves, and even lower, to ways in which the chemical composition of leaves, roots, and organic matter can be determined analytically. In other words, there are numerous processes, arranged in more of a level of resolution fashion (Fig. 1.5), which can be studied in a comparative fashion. We most often compare processes between stands of vegetation or across watersheds, particularly when following integrating variables such as water flow, which influences nutrient and particulate organic matter and dissolved organic matter exports. If we were more concerned with air pollutants or gaseous losses of NO_x or CH_4, we would work at a more regional or airshed level of resolution (Haines and Swank, 1988).

BIODIVERSITY AND SOIL ECOLOGY

There is a rapidly growing interest among biologists in the fates of the very diverse array of organisms in all ecosystems of the world. What do we know of the full species richness to make even any educated guesses about the total extent of the organisms or how many of them may be in an endangered status (Hawksworth, 1991a; Coleman et al., 1994b)? We present two examples in point, taken from Coleman et al. (1994b). There are currently 70,000 species of fungi described (Table 7.2). By assuming that a constant ratio of species of fungi exists to those plant species already known, Hawksworth (1991b) calculated that there may be a total of 1.5 million species of fungi described when we are all finished with this mammoth classification task.

In an example using soil fauna, the oribatid mites (see chapter 4), many of which are fungal and detrital feeders, have been fairly well studied in the Northern Hemisphere. However, only 30–35% of the oribatids in North America have been adequately described (Behan-Pelletier and Bissett, 1993), despite many efforts carried out since the 1960s. They suggest that there may be more than 100,000 undescribed species of oribatid mites yet to be discovered. Particularly in many tropical regions, oribatids and other small arthropods are not well known in both soil and tree canopy environments (V. Behan-Pelletier, personal communication, 1994). This difficulty is

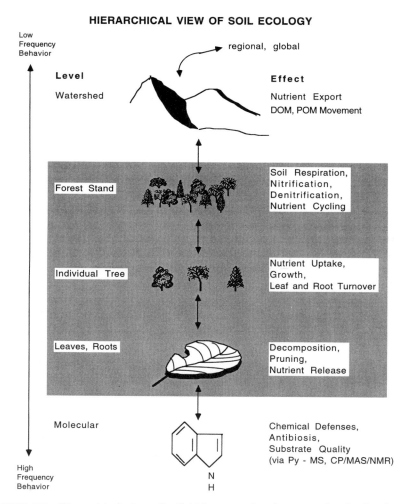

FIGURE 7.5 Hierarchical view of soil biology ranging from a molecular level up to watersheds and beyond. The soil-related levels of resolution are stippled to highlight the central organizing function of soil processes. In some cases, feedbacks from vegetation and soil interactions reverberate back down to the molecular level, and flows should be viewed as potentially going both ways. DOM, dissolved organic matter; POM, particulate organic matter; PY-MS, pyrolysis mass spectrometry; CP, cross-polarized; MAS, magic angle spinning; NMR, nuclear magnetic resonance (Reprinted with permission from Coleman *et al.*, 1992. Copyright 1992 Springer-Verlag.).

compounded by our very poor knowledge of identities of the immature stages of soil fauna, particularly the Acari. A solution to this problem may require the considerable application of molecular techniques to more effectively work with all life stages of the soil fauna (Coleman,

TABLE 7.2.   Comparison of the Numbers of Known and Estimated Total Species Globally of Selected Groups or Organisms[a]

Group	Known species	Estimated total species	Percentage known
Vascular plants	220,000	270,000	81
Bryophytes	17,000	25,000	68
Algae	40,000	60,000	67
Fungi	69,000	1,500,000	5
Bacteria	3,000	30,000	10
Viruses	5,000	130,000	4

[a]From Hawksworth (1991b).

1994a; Freckman, 1994). We concur totally with the following comment from Behan-Pelletier and Bissett (1993): "(A)dvances in systematics and ecology must progress in tandem: systematics providing both the basis and predictions for ecological studies, and ecology providing information on community structure and explanations for recent evolution and adaptation."

What are the consequences of biodiversity? Does the massive array of hundreds of thousands of fungi and perhaps millions of bacterial species make sense in any ecological or evolutionary context? What is the implication of the apparent "excess" of species diversity of soil microflora, where many species exist at a very low frequency and in an inactive state? If we maintain a large species richness and accompanying large genetic pools in soils, what is its impact on the evolution of new taxa? What are the implications for ecosystem function if this degree of redundancy exists; does it imply that some of the organisms are somehow vestigial remnants or relics of bygone conditions (Coleman *et al.*, 1994b)? What are the functional roles of such hidden or apparently cryptic organisms, are they performing some essential, but totally unknown functions, perhaps at microsites which we do not observe or work with? One approach which may show promise is the use of reporter genes linked to gene promoters in order to measure, *in situ*, the activity of specific enzymes related to defined processes (Wilson *et al.*, 1994). We need to link specific methods such as noted here with soil thin-section studies, such as those of Tippkötter *et al.* (1986), Postma and Altemüller (1990), and Foster (1994). Such means will enable the inclusion of spatial dimensions to soil ecological studies; the addition of temporal ones provides the much-needed aspect of time as well. Soils are rife with historical signs and "fingerprints," as has been made evident by studies using radiotracers and stable isotopes (Stout *et al.*, 1981; Nadelhoffer *et al.*, 1985).

What are the linkages between biodiversity and ecosystem function? It should be possible to look for natural "experiments," such as regions with low species richness, e.g., on an island, versus sites at similar latitudes, which are on continents, where key ecosystem processes can be measured, such as rates of decomposition or nutrient cycling. Studies of the interactions of species richness and ecosystem processes are reviewed masterfully by Vitousek (1994). Under such conditions, all of the major abiotic factors are held reasonably similar, allowing for the study of the impacts of species richnesses of key indicator microflora or fauna on ecosystem processes of interest. Such experiments are certainly performable and may yield some surprising results.

SOILS AND "GAIA": POSSIBLE MECHANISMS FOR EVOLUTION OF "THE FITNESS OF THE SOIL ENVIRONMENT"?

As was mentioned in Chapter 1, there are many positive feedback mechanisms in soils in which organisms have arisen and/or evolved together. These include roots and arbuscular mycorrhiza (AM) and many of the genera and families of soil fauna. The following discussion is based on the very insightful and stimulating article by van Breemen (1992) entitled "Soils: Biotic Constructions in a Gaian sense? van Breemen asks the central question: have soils merely been influenced by biota or have biota created soils as natural bodies with properties favorable for terrestrial life? He presents hypotheses or postulates related to the overarching theme: (1) there are soil properties "favorable" for terrestrial life in general; (2) biota, including plants and the soil-dwelling organisms, are able to affect those soil properties; 3) on a scale of ecosystems and a global ("Gaian") scale , biotic action makes the outermost (1–100 cm) layer of the earth's crust more favorable for terrestrial life in general than it would have been in their absence; (4) at an ecosystem-level scale, biota tend to offset the effects of unfavorable properties of the soil or soil parent material by modifying those soil properties; and (5) modification of soil properties may play a role in species competition.

Following from the ideas of Odum and Biever (1984), there should be some positive or donor recipient controls on interactions between primary producers and other biota; the AM is a prominent example. As noted earlier, feeding on detrital organic matter in the soil is generally the principal energy flow in terrestrial ecosystems. Therefore, feedback loops arising in the soil community (such as detrital food webs, Chapter 6) should have a major effect on net primary productivity. Thus, soil–biota interactions may be a most fruitful area to investigate and test hypotheses about positive effects of biota on the environment.

For favorable soil properties, changes and general improvements in soil porosities and soil organic matter status are prime examples of general improvements in soil characteristics by cumulative interactions of the soil biota. This is not a simple linear progression, however; there are examples of surface-feeding earthworms that remove enough of the surface leaf litter material that there is indeed an accelerated amount of soil erosion in their presence than in their absence (Johnson, 1990).

In the areas of soil texture and structure, as well as soil chemical properties, there are numerous examples of soil–biotic interactions having a generally beneficial effect in the top meter of soil material. One example of this is provided by Gill and Abrol (1986) who described how planting *Eucalyptus teretocornis* and *Acacia nilotica* on an alkali soil (pH 10.5) markedly decreased pH and salinity within 3 to 6 years. These changes were probably caused by a suite of factors, including increased water permeability, following the development of root channels and the accumulation of organic matter in the upper 20–50 cm of the soil profile. Other biota, notably termites, can promote higher salt content in soils, as detected by measurements in inhabited and abandoned termite hills compared to the surrounding soil (De Wit, 1978). Many of these processes tend to increase the amounts of heterogeneity within soil profiles, which has been well reviewed by Stark (1994).

At both ecosystem and global scales, there are significant effects of biota on rock and soil weathering. The early pioneering researches of Vernadsky (1944) and Volobuev (1964) in particular originated and made popular the concept of "organic weathering." The able partnership of roots and microbes in mineral translocation is noteworthy, e.g., removing the interlayer K from phlogopite (vermiculitization) within the first 2 mm of the rhizosphere. Other references on biological impacts on mineral weathering can be found in Schlesinger (1991). As noted in Chapter 6, the soil physical effects of earthworms on soil structure, formation of heterogeneous pores, and high structural stability are hallmarks of soil–biota interactions over long time intervals.

There are several examples of transformations which counteract unfavorable soil properties. These arise principally from the influence of the biota on translocation and concentration of nutrients in the upper 1 m of the earth's mantle, the living soil. In general, biota tend to invest more in increasing nutrient supply under nutrient-poor than in nutrient-rich conditions. Root production and activity, as a fraction of total net primary production, tend to be higher in the nutrient-poor conditions (Odum, 1971). This should be considered against a background of the generally slow growth rates and nutrient fluxes which occur in many wild plants on low nutrient soils (Chapin, 1980). There is also an intriguing nutrient conservation process which occurs in many low nutrient ecosystems. The development of mor humus types, characterized by thick organic horizons, is typical for "poor" (low productivity) sites and may represent nutrient conservation brought about as a result of the poorly decomposable litter formed in the surface layers (Vos and Stortelder, 1988). This in turn may lead to a further inhibition of decomposition and net primary production; this is an example of a positive feedback effect, which may require occasional fires or other disturbances to act as a suitable "reset" over millennial time spans. Soil phosphorus, in its various inorganic and organic forms, is perhaps the most limiting element in terrestrial ecosystems (van Breemen, 1992) The storage of phosphorus by secondary iron and aluminum phases is partly under biotic control and may be regarded as part of the tight biotic cycling of phosphorus for three reasons: (1) secondary iron and aluminum oxides result from the biologically mediated weathering of primary minerals; (2) secondary iron and aluminum oxides are often precipitated under the influence of iron oxidizing bacteria; and (3) the oxides can be kept in a mostly amorphous form by interaction with humic substances (Guillet, 1990), from which phosphorus can be extracted by plants more efficiently than from crystalline oxides.

All of the foregoing perhaps raises more questions than answers. However, the general trend is for the number of species and individuals with positive effects to increase, both in successional sequences and over evolutionary time. In essence, the property of an individual that improves the environment for that individual or increases its reproductive success will benefit both it and its competitors as well. The selective advantage for such a trait(s) is probably small, viewed in a classical Darwinian context. If viewed in more general contexts, such as enhancement of site qualities, then this can be considered a more general application of community and ecosystem development. van Breemen (1992) notes that development of a trait in an earthworm, allowing it to better control the moisture content and CO_2/O_2 balance of its immediate surroundings, would benefit other organisms and site properties. If requirements of plants or a plant species happen to match those of the earthworm, then coevolution of the plant and worm might be possible too. Wilson (1980) suggested that one might envision further development and evolution of a community of microbes, which could coevolve with the earthworm, to better enhance nutrient cycling processes. This is an evolutionary example of significant processes at "hot spots," as noted in Chapter 6. The scenario is speculative, but serves as an example of where we may expect to see additional breakthroughs occurring in the cryptic and fascinating world of soil ecology.

References

Albers, B. P., Beese, F., and Hartman, A. (1995). Flow-microcalorimetry measurements of aerobic and anaerobic soil microbial activity. *Biol. Fertil. Soils* **19**, 203–208.

Allen, M. F. (1991). "The Ecology of Mycorrhizae." Cambridge Univ. Press, Cambridge.

Allen, M. F., ed. (1992). "Mycorrhizal Functioning: An Integrative Plant–Fungal Process." Chapman & Hall, London.

Allen, T. F. H., and Starr, T. B. (1982). "Hierarchy. Perspectives for Ecological Complexity." Univ. Chicago Press, Chicago, Illinois.

Allen-Morley, C. R., and Coleman, D. C. (1989). Resilience of soil biota in various food webs to freezing perturbations. *Ecology* **70**, 1127–1141.

Alphei, J., Bonkowski, M., and Scheu, S. (1995). Application of the selective inhibition method to determine bacterial:fungal ratios in three beechwood soils rich in carbon-optimization of inhibitor concentrations. *Biol. Fertil. Soils* **19**, 173–176.

Anderson, D. W. (1988). The effect of parent material and soil development on nutrient cycling in temperate ecosystems. *Biogeochemistry* **5**, 71–97.

Anderson, D. W., and Coleman, D. C. (1985). The dynamics of organic matter in grassland soils. *J. Soil Water Conserv.* **40**, 211–216.

Anderson, J. M. (1975). Succession, diversity and trophic relationships of some soil animals in decomposing leaf litter. *J. Anim. Ecol.* **44**, 475–495.

Anderson, J. M. (1992). Responses of soils to climate change. *Adv. Ecol. Res.* **22**, 163–210.

Anderson, J. M., Huish, S. A., Ineson, P., Leonard, M. A., and Splatt, P. R. (1985). Interactions of invertebrates, micro-organisms, and tree roots in nitrogen and mineral element fluxes in deciduous woodland soils. *In* "Ecological Interactions in Soil: Plants, Microbes and Animals" (A. H. Fitter, D. Atkinson, D. J. Read, and M. B. Usher, eds.), pp. 377–392. British Ecological Society Special Publication 4, Oxford.

Anderson, J. P. E., and Domsch, K. H. (1978). A physiological method for the quantitative measurement of microbial biomass in soils. *Soil Biol. Biochem.* **10**, 215–221.

Anderson, R. V., Coleman, D. C., and Cole, C. V. (1981a). Effects of saprotrophic grazing on net mineralization. Terrestrial nitrogen cycles. *Ecol. Bull.* **33**, 201–216.

Anderson, R. V., Coleman, D. C., Cole, C. V., and Elliott, E. T. (1981b). Effect of the nematodes *Acrobeloides* sp. and *Mesodiplogaster lheritieri* on substrate utilization and nitrogen and phosphorus mineralization in soil. *Ecology* **62**, 549–555.

Anderson, R. V., Gould, W. D., Woods, L. E., Cambardella, C., Ingham, R. E., and Coleman, D. C. (1983). Organic and inorganic nitrogenous losses by microbivorous nematodes in soil. *Oikos* **40**, 75–80.

Anderson, T.-H. (1994). Physiological analysis of microbial communities in soil: Applications and limitations. *In* "Beyond the Biomass" (K. Ritz, J. Dighton, and K. E. Giller, eds.), pp. 67–76. Wiley, Chichester.

Anderson, T.-H., and Domsch, K. H. (1993). The metabolic quotient for CO_2 (qCO_2) as a specific activity parameter to assess the effects of environmental conditions, such as pH, on the microbial biomass of forest soils. *Soil Biol. Biochem.* **25**, 393–395.

Andrén, O., Lindberg, T., Paustian, K., and Rosswall, T., eds. (1990). "Ecology of Arable Land. Organisms, Carbon and Nitrogen Cycling." Munksgaard, Copenhagen.

Arlian, L. G., and Woolley, T. A. (1970). Observations on the biology of *Liacarus cidarus* (Acari: Cryptostigmata, Liacaridae). *J. Kans. Entomol. Soc.* **43**, 297–301.

Arnett, R. H., Jr. (1993). "American Insects: A Handbook of the Insects of America North of Mexico." Sandhill Crane Press, Gainesville, Florida.

Arrhenius, S. (1896). On the influence of carbolic acid in the air upon the temperature of the ground. *Philos. Mag.* **41**, 237–257.

Bååth, E., Lohm, U., Lundgren, B., Rosswall, T., Soderstrom, B., and Sohlenius, B. (1981). Impact of microbial-feeding animals on total soil activity and nitrogen dynamics: A soil microcosm experiment. *Oikos* **37**, 257–264.

Baker, E. W., and Wharton, G. W. (1952). "Introduction to Acarology." MacMillan, New York.

Baker, E. W., Camin, J. H., Cunliffe, F., Woolley, T. A., and Yunker, C. E. (1958). "Guide to the Families of Mites." Institute of Acarology, University of Maryland, College Park.

Bal, L. (1982). "Zoological Ripening of Soils." Pudoc, Wageningen.

Bamforth, S. S. (1980). Terrestrial protozoa. *J. Protozool.* **27**, 33–36.

Barber, D. A., and Martin, J. K. (1976). The release of organic substances by cereal roots into soil. *New Phytol.* **76**, 69–80.

Barra, J. A., and Christiansen, K. (1975). Experimental study of aggregation during the development of *Psuedosinella impediens* (Collembola, Entomobryidae). *Pedobiologia* **15**, 343–347.

Battley, E. H. (1987). "Energetics of Microbial Growth." Wiley (Interscience), New York.

Beare, M. H., Neely, C. L., Coleman, D. C., and Hargrove, W. L. (1990). A substrate-induced respiration (SIR) method for measurement of fungal and bacterial biomass on plant residues. *Soil Biol. Biochem.* **22**, 585–594.

Beare, M. H., Neely, C. L., Coleman, D. C., and Hargrove, W. L. (1991). Characterizations of a substrate-induced respiration method for measuring fungal, bacterial and total microbial biomass on plant residues. *Agric. Ecosyst. Environ.* **34**, 65–73.

Beare, M. H., Parmelee, R. W., Hendrix, P. F., Cheng, W., Coleman, D. C., and Crossley, D. A., Jr. (1992). Microbial and faunal interactions and effects on litter nitrogen and decomposition in agroecosystems. *Ecol. Monogr.* **62**, 569–591.

Beare, M. H., Hendrix, P. F., and Coleman, D. C. (1994a). Water-stable aggregates and organic matter fractions in conventional and no-tillage soils. *Soil Sci. Soc. Am. J.* **58**, 777–786.

Beare, M. H., Cabrera, M. L., Hendrix, P. F., and Coleman, D. C. (1994b). Aggregate-protected and unprotected pools of organic matter in conventional and no-tillage ultisols. *Soil Sci. Soc. Am. J.* **58**, 787–795.

Beare, M. H., Coleman, D. C., Crossley, D. A., Jr., Hendrix, P. F., and Odum, E. P. (1995). A hierarchical approach to evaluating the significance of soil biodiversity to biogeochemical cycling. *Plant Soil* **170**, 5–22.

Behan, V. M. and Hill, S. B. (1978). Feeding habits and spore dispersal of oribatid mites in the North American arctic. *Rev. Ecol. Biol. Sol* **15**, 497–516.

Behan-Pelletier, V. M. (1993). Eremaeidae (Acari:Oribatida) of North America, *Memoirs, Ent. Soc. Canada,* No 168, pp. 193.

Behan-Pelletier, V. M., and Bissett, B. (1993). Biodiversity of nearctic soil arthropods. *Can. Biodiv.* **2**, 5–14.

Behan-Pelletier, V. M., and Hill, S. B. (1983). Feeding habits of sixteen species of Oribatei (Acari) from an acid peat bog, Glenamoy, Ireland. *Rev. Ecol. Biol. Sol* **20**, 221–267.

Bentham, H., Harris, J. A., Birch, P., and Short, K. C. (1992). Habitat classification and soil restoration assessment using analysis of soil microbiological and physico-chemical characteristics. *J. Appl. Ecol.* **29**, 711–718.

Berg, B. and Staaf, H. (1981). Leaching, accumulation and release of nitrogen in decomposing forest litter. *Ecol. Bull.* **33**, 163–178.

Bignell, D. E. (1984). The arthropod gut as an environment for microorganisms. *In* "Invertebrate–microbial Interactions" (J. M. Anderson, A. D. M. Rayner, and D. W. H. Walton, eds.), pp. 205–227. Cambridge Univ. Press, Cambridge.

Blair, J. M. (1988a). Nitrogen, sulfur and phosphorus dynamics in decomposing deciduous leaf litter in the southern Appalachians. *Soil Biol. Biochem.* **20**, 693–701.

Blair, J. M. (1988b). Nutrient release from decomposing foliar litter of three tree species with special reference to calcium, magnesium and potassium dynamics. *Plant Soil* **110**, 49–55.

Blair, J. M., and Crossley, D. A., Jr. (1988). Litter decomposition, nitrogen dynamics and litter microarthropods in a southern Appalachian hardwood forest 8 years following clearcutting. *J. Appl. Ecol.* **25**, 683–698.

Blair, J. M., Crossley, D. A., Jr., and Callaham, L. C. (1992). Effects of litter quality and microarthropods on N dynamics and retention of exogenous 15N in decomposing litter. *Biol. Fertil. Soils* **12**, 241–252.

Böhm, W. (1979). "Methods of Studying Root Systems." Springer-Verlag, Berlin.

Bongers, T. (1990). The maturity index: An ecological measure of environmental disturbance based on nematode species composition. *Oecologia* **83**, 14–19.

Bornebusch, C. H. (1930). "The Fauna of Forest Soil." Copenhagen.

Borror, D. J., DeLong, D. M., and Triplehorn, C. A. (1981). "An Introduction to the Study of Insects," 5th Ed. Saunders College, Philadelphia.

Bouché, M. B. (1975). Action de la faune sur les états de la matiére organique dans les écosystemes. *In* "Biodégradation et Humification" (G. Kilbertus, O. Reisinger, A. Mourey, and J. A. Cancela da Fonseca, eds.), pp. 157–168. Pierron, Sarrugemines, France.

Bouché, M. B. (1977). Stratégies lombriciennes. *Ecol. Bull.* **25**, 122–132.

Bouwman, A. F., ed. (1990). "Soils and the Greenhouse Effect." Wiley, Chichester.

Bowen, H. J. M. (1979). "Environmental Chemistry of the Elements." Academic Press, London and New York.

Box, J. E., Jr., and Hammond, L. C. (1990). "Rhizosphere Dynamics." Westview Press, Boulder, Colorado.

Box, J. E., Jr., and Johnson, J. W. (1987). Minirhizotron rooting comparisons of three wheat cultivars. *In* "Minirhizotron Observation Tubes: Methods and Applications for Measuring Rhizosphere Dynamics" (H. M. Taylor, ed.), pp. 123–130. American Society of Agronomy Special Publication No. 50, Madison, Wisconsin.

Brady, N. C. (1974). "The Nature and Properties of Soils," 8th Ed. MacMillan, New York.

Breznak, J. A. (1984). Biochemical aspects of symbiosis between termites and their intestinal microbiota. *In* "Invertebrate–Microbial Interactions" (J. M. Anderson, A. D. M. Rayner, and D. W. H. Walton, eds.), pp. 173–203. Cambridge Univ. Press, Cambridge.

Brussaard, L., and Kooistra, M. J. (1993). "Soil Structure/Soil Biota Interrelationships." Elsevier, Amsterdam.

Bryant, R. J., Woods, L. E., Coleman, D. C., Fairbanks, B. C., McClellan, J. F., and Cole, C. V. (1982). Interactions of bacterial and amoebal populations in soil microcosms with fluctuating moisture content. *Appl. Environ. Microbiol.* **43**, 7447–752.

Buckman, H. O., and Brady, N. C. (1970). "The Nature and Properties of Soils," 7th Ed. Macmillan, New York.

Caldwell, M. M., and Camp, L. B. (1974). Belowground productivity of two cool desert communities. *Oecologia* **17**, 123–130.

Cambardella, C. A., and Elliott, E. T. (1994). Carbon and nitrogen dynamics of soil organic matter fractions from cultivated grassland soils. *Soil Sci. Soc. Am. J.* **58**, 123–130.

Chakraborty, S., and Warcup, J. H. (1983). Soil amoebae and saprophytic survival of *Gaeumannomyces graminis tritici* in a suppressive pasture soil. *Soil Biol. Biochem.* **15**, 181–185.

Chakraborty, S., Old, K. M., and Warcup, J. H. (1983). Amoebae from a take-all suppressive soil which feed on *Gaeumannomyces graminis tritici* and other soil fungi. *Soil Biol. Biochem.* **15**, 17–24.

Chapin, F. S., III. (1980). The mineral nutrition of wild plants. *Ann. Rev. Ecol. System.* **11**, 233–260.

Cheng, W., and Coleman, D. C. (1990). Effect of living roots on soil organic matter decomposition. *Soil Biol. Biochem.* **22**, 781–787.

Cheng, W., Coleman, D. C., and Box, J. E., Jr. (1990). Root dynamics, production and distribution in agroecosystems on the Georgia piedmont using minirhizotrons. *J. Appl. Ecol.* **27**, 592–604.

Cheng, W., Coleman, D. C., Carroll, C. R., and Hoffman, C. A. (1993). *In situ* measurement of root respiration and soluble carbon concentrations in the rhizosphere. *Soil Biol. Biochem.* **25**, 1189–1196.

Cheshire, M. (1979). "Soil Carbohydrates." Academic Press, London.

Cheshire, M. V., Sparling, G. P., and Mundie, C. M. (1984). Influence of soil type, crop and air drying on residual carbohydrate content and aggregate stability after treatment with periodate and tetraborate. *Plant Soil* **76**, 339–347.

Chiariello, N., Hickman, J. C., and Mooney, H. A. (1982). Endomycorrhizal role for interspecific transfer of phosphorus in a community of annual plants. *Science* **217**, 941–943.

Christiansen, K. (1970). Experimental studies on the aggregation and dispersion of Collembola. *Pedobiologia* **10**, 180–198.

Christiansen, K. A. (1990). Insecta: Collembola. *In* "Soil Biology Guide" (D. L. Dindal, ed.), pp. 965–995. Wiley, New York.

Christiansen, K. A., and Bellinger, P. F. (1980–1981). "The Collembola of North America North of the Rio Grande." Grinnell College, Grinnell, Iowa.

Christiansen, K., Doyle, M., Kahlert, M., and Gobaleza, D. (1992). Interspecific interactions between collembolan populations in culture. *Pedobiologia* **36**, 274–286.

Christie, J. R., and Perry, V. G. (1951). Removing nematodes from soil. *Proc. Helminthol. Soc. Wash.,* 106–108.

Clarholm, M. (1981). Protozoan grazing of bacteria in soil-impact and importance. *Microb. Ecol.* **7**, 343–350.

Clarholm, M. (1985). Possible roles for roots, bacteria, protozoa and fungi in supplying nitrogen to plants. *In* "Ecological Interactions in Soil: Plants, Microbes and Animals" (A. H. Fitter, D. Atkinson, D. J. Read, and M. B. Usher, eds.), pp. 355–365. Blackwell, Oxford.

Clarholm, M. (1994). The microbial loop in soil. *In* "Beyond the Biomass" (K. Ritz, J. Dighton, and K. E. Giller, eds.), pp. 221–230. Wiley, Chichester.

Cole, L. C. (1946). The cryptozoa of an Illinois woodland. *Ecol. Monogr.* **16**, 49–86.

Coleman, D. C. (1976). A review of root production processes and their influence on soil biota in terrestrial ecosystems. *In* "The Role of Terrestrial and Aquatic Organisms in Decomposition Processes" (J. M. Anderson and A. Macfadyen, eds.), pp. 417–434. Blackwell, Oxford.

Coleman, D. C. (1985). Through a ped darkly: An ecological assessment of root–soil–microbial–faunal interactions. *In* "British Ecological Society Special Publication Number 4" (A. H. Fitter, D. Atkinson, D. J. Read, and M. B. Usher, eds.), pp. 1–21. Blackwell, Oxford.

Coleman, D. C. (1994a). Compositional analysis of microbial communities. *In* "Beyond the Biomass" (K. Ritz, J. Dighton, and K. Giller, eds.), pp. 201–220. Wiley, Chichester.

Coleman, D. C. (1994b). The microbial loop concept as used in terrestrial soil ecology studies. *Microb. Ecol.* **28**, 245–250.

Coleman, D. C., and Hendrix, P. F. (1988). Agroecosystems Processes. *In* "Concepts of Ecosystem Ecology, a Comparative View" (L. R. Pomeroy and J. Alberts, eds.), pp. 149–170. Springer-Verlag, New York.

Coleman, D. C., and Sasson, A. (1980). Decomposers subsystem. *In* "Grasslands, Systems Analysis, and Man. IBP Synthesis Volume 19" (A. Breymeyer and G. van Dyne, eds.), Chapter 7, pp. 609–655. Cambridge Univ. Press, London.

Coleman, D. C., and Schoute, J. F. T. (1993). Translation of soil features across levels of spatial resolution—Introduction to round table discussion. *Geoderma* **57**, 171–181.

Coleman, D. C., Andrews, R., Ellis, J. E., and Singh, J. S. (1976). Energy flow and partitioning in selected man-managed and natural ecosystems. *Agro-Ecosystems* **3**, 45–54.

Coleman, D. C., Cole, C. V., Anderson, R. V., Blaha, M., Campion, M. K., Clarholm, M., Elliott, E. T., Hunt, H. W., Schaefer, B., and Sinclair, J. (1977). Analysis of rhizosphere–saprophage interactions in terrestrial ecosystems. *Ecol. Bull.* **25**, 299–309.

Coleman, D. C., Reid, C. P. P., and Cole, C. V. (1983). Biological strategies of nutrient cycling in soil systems. *Adv. Ecol. Res.* **13**, 1–55.

Coleman, D. C., Cole, C. V., and Elliott, E. T. (1984). Decomposition, organic matter turnover and nutrient dynamics in agroecosystems. *In* "Agricultural Ecosystems— Unifying Concepts" (R. Lowrance, B. R. Stinner, and G. J. House, eds.), pp. 83–104. Wiley (Interscience), New York.

Coleman, D. C., Ingham, E. R., Hunt, H. W., Elliott, E. T., Reid, C. P. P., and Moore, J. C. (1990). Seasonal and faunal effects on decomposition in semiarid prairie, meadow and lodgepole pine forest. *Pedobiologia* **34**, 207–219.

Coleman, D. C., Odum, E. P., and Crossley, D. A., Jr. (1992). Soil biology, soil ecology, and global change. *Biol. Fertil. Soils* **14**, 104–111.

Coleman, D. C., Hendrix, P. F., Beare, M. H., Cheng, W., and Crossley, D. A., Jr. (1993). Microbial and faunal dynamics as they affect soil organic matter dynamics in subtropical Agroecosystems. *In* "Soil Biota and Nutrient Cycling Farming Systems" (M. G. Paoletti, W. Foissner, and D. C. Coleman, eds.), pp. 1–14. Lewis, Chelsea, Michigan.

Coleman, D. C., Hendrix, P. F., Beare, M. H., Crossley, D. A., Jr., Hu, S., and van Vliet, P. C. J. (1994a). The impacts of management and biota on nutrient dynamics and soil structure in sub-tropical agroecosystems: Impacts on detritus food webs. *In* "Soil Biota Management in Sustainable Farming Systems" (C. E. Pankhurst, B. M. Doube, V. V. S. R. Gupta, and P. R. Grace, eds.), pp. 133–143. CSIRO, Melbourne, Australia.

Coleman, D. C., Dighton, J., Ritz, K., and Giller, K. E. (1994b). Perspectives on the compositional and functional analysis of soil communities. *In* "Beyond the Biomass" (K. Ritz, J. Dighton, and K. E. Giller, eds.), pp. 261–271. Wiley, Chichester.

Coûteaux, M.-M. (1972). Distribution des thécamoebiens de la litière et de l'humus de deux sols forestier d'humas brut. *Pedobiologia* **12**, 237–243.

Coûteaux, M.-M. (1985). Relation entre la densité apparente d'un humus et l'aptitude à la croissance de ses ciliés. *Pedobiologia* **28**, 289–303.

Crawford, C. S. (1981). "Biology of Desert Invertebrates." Springer-Verlag, New York.

Crocker, R. L. (1952). Soil genesis and the pedogenic factors. *Q. Rev. Biol* **27**, 139–168.

Cromack, K. (1973). "Litter production and litter decomposition in a mixed hardwood watershed and in a white pine watershed at Coweeta Hydrologic Station, North Carolina." Ph.D. Dissertation, University of Georgia, Athens.

Crossley, D. A. Jr., and Hoglund, M. P. (1962). A litter-bag method for the study of microarthropods inhabiting leaf litter. *Ecology* **43**, 571–573.

Crossley, D. A., Jr., Mueller, B. R., and Perdue, J. C. (1992). Biodiversity of microarthropods in agricultural soils: Relations to functions. *Agric. Ecosyst. Environ.* **40**, 37–46.

Crowe, J. H. (1975). The physiology of cryptobiosis in tardigrades. *Mem. Ist. Ital. Idrobiol.* **32** (Suppl.), 37–59.

Crowe, J. H., and Cooper, A. F., Jr. (1971). Cryptobiosis. *Sci. Am.* **225**, 30–36.

Curl, E. A. (1979). Effects of mycophagous collembola on *Rhizoctonia solani* and cotton-seedling disease. *In* "Soil-Borne Plant Pathogens" (B. Schippers and W. Gams, eds.), pp. 253–269. Academic Press, London.

Curl, E. A., and Truelove, B. (1986). "The Rhizosphere." Springer-Verlag, Berlin, New York.

Curry, J. P., Byrne, D., and Boyle, K. E. (1995). The earthworm population of a winter cereal field and its effects on soil and nitrogen turnover. *Biol. Fertil. Soils* **19**, 166–172.

Cutler, D. W. (1920). A method for estimating the number of active protozoa in the soil. *J. Agric. Sci.* **10**, 135–143.

Cutler, D. W. (1923). The action of protozoa on bacteria when inoculated into sterile soil. *Ann. Appl. Biol.* **10**, 137–141.

Cutler, D. W., Crump, L. M., and Sandon, H. (1923). A quantitative investigation of the bacterial and protozoan population of the soil. *Philos. Trans. R. Soc. B Biol. Sci.* **211**, 317–350.

Daniel, O., and Anderson, J. M. (1992). Microbial biomass and activity in contrasting soil materials after passage through the gut of the earthworm *Lumbricus rubellus* Hoffmeister. *Soil Biol. Biochem.* **24**, 465–470.

Darbyshire, J. F. (1994). "Soil Protozoa." CAB International, Wallingford, U.K.

Darbyshire, J. F., and Greaves, M. (1967). Protozoa and bacteria in the rhizosphere of *Sinapis alba* (L), *Trifolium repens* (L.), and *Lolium perenne* (L). *Can. J. Microbiol.* **13**, 1057–1068.

Darwin, C. (1881). "The Formation of Vegetable Mould, through the Action of Worms, with Observations on Their Habits." Murray, London.

Dash, M. C. (1990). Oligochaeta: Enchytraeidae. *In* "Soil Biology Guide" (D. L. Dindal, ed.), pp. 311–340. Wiley, New York.

DeAngelis, D. L. (1992). "Dynamics of Nutrient Cycling and Food Webs." Chapman & Hall, London.

De Ruiter, P. C., Moore, J. C., Zwart, K. B., Bouwman, L. A., Hassink, J., Bloem, J., De Vos, J. A., Marinissen, J. C. Y., Didden, W. A. M., Lebbink, G., and Brussaard, L. (1993). Simulation of nitrogen mineralization in the belowground food webs of two winter wheat fields. *J. Appl. Ecol.* **30**, 95–106.

De Ruiter, P. C., Neutel, A.-M., and Moore, J. C. (1995). Energetics and stability in belowground food webs. *In* "Integration of Patterns and Dynamics" (G. A. Polis and K. O. Winemiller, eds.), in press. Chapman & Hall, New York.

De Wit, H. A. (1978). Soils and grassland types of the Serengeti plain (Tanzania). Ph.D. Thesis. Agricultural University Wageningen, The Netherlands.

Didden, W. A. M. (1990). Involvement of Enchytraeidae (Oligochaeta) in soil structure evolution in agricultural fields. *Biol. Fertil. Soils* **9**, 152–158.

Dighton, J. and Coleman, D. C. (1992). Phosphorus relations of roots and mycorrhizas of *Rhododendron maximum* L. in the Southern Appalachians, N. Carolina. *Mycorrhiza* **1**, 175–184.

Dindal, D. L. (1990). "Soil Biology Guide." Wiley, New York.

Donner, J. (1966). "Rotifers." Warne, London.

Doran, J. W. (1980a). Microbial changes associated with residue management with reduced tillage. *Soil Sci. Soc. Am. J.* **44**, 518–524.

Doran, J. W. (1980b). Soil microbial and biochemical changes associated with reduced tillage. *Soil Sci. Soc. Am. J.* **44**, 765–771.

Doran, J. W., Coleman, D. C., Bezdicek, D. F., and Stewart, B. A., eds. (1994). "Defining Soil Quality for a Sustainable Environment." SSSA Special Publication American Society of Agronomy, Madison, Wisconsin.

Dunger, W. (1983). "Fauna in Soils." Auflagen Ziemsen Verlag, Wittenberg Lutherstadt, Germany.

Eash, N. S., Karlen, D. L., and Parkin, T. B. (1994). Fungal contributions to soil aggregation and soil quality. *In* "Defining Soil Quality for a Sustainable Environment" (J. Doran, D. C. Coleman, D. F. Bezdicek, and B. A. Stewart, eds.), pp. 221–228. SSSA Special Publication No. 35, Madison, Wisconsin.

Edwards, C. A. (1990). Symphyla. *In* "Soil Biology Guide" (D. L. Dindal, ed.), pp. 891–910. Wiley (Interscience), New York.

Edwards, C. A. (1991). The assessment of populations of soil-inhabiting invertebrates. *Agric. Ecosyst. Environ.* **34**, 145–176.

Edwards, C. A., and Bohlen, P. (1995). "Biology of Earthworms," 3rd Ed., Chapman & Hall, London.

Edwards, C. A, and Dennis, E. B. (1962). The sampling and extraction of Sylmphyla from soil. *In* "Progress in Soil Zoology." (P. W. Murphy, ed.), pp. 300–304. Butterworth, London.

Edwards, C. A, and Heath, G. W. (1963). The role of soil animals in breakdown of leaf material. *In* "Soil Organisms" (J. Doeksen and J. van der Drift, eds.), pp. 76–80. North-Holland Publ., Amsterdam.

Edwards, C. A., and Lofty, J. R. (1977). "Biology of Earthworms," 2nd Ed." Chapman & Hall, London.

Eisenbeis, G., and Wichard, W. (1987). "Atlas on the Biology of Soil Arthropods." Springer-Verlag, Stuttgart.

Elliott, E. T. (1986). Hierarchic aggregate structure and organic C, N, and P in native and cultivated grassland soils. *Soil Sci. Soc. Am. J.* **50**, 627–633.

Elliott, E. T. (1994). Embodying process information in models evaluated with site network information: Nairobi Workshop. *Trans. 15th World Congr. Soil Sci., INEGI & CNA*, 163–177.

Elliott, E. T., and Coleman, D. C. (1977). Soil protozoan dynamics in a shortgrass prairie. *Soil Biol. Biochem.* **9**, 113–118.

Elliott, E. T., and Coleman, D. C. (1988). Let the soil work for us. *Ecol. Bull.* **39**, 2332.

Elliott, E. T., Janzen, H. H., Campbell, C. A., Cole, C. V., and Myers, R. J. K. (1994). Principles of ecosystem analysis and their application to integrated nutrient management and assessment of sustainability, *Proc. Sustainable Land Management 21st Century. Vol. 2: Plenary Papers*. Can. Soc. Soil Sci., Lethbridge, Alta., Canada.

Elliott, E. T., Horton, K., Moore, J. C., Coleman, D. C., and Cole, C. V. (1984). Mineralization dynamics in fallow dryland wheat plots, Colorado. *Plant Soil* **76**, 149–155.

Elton, C. S. (1927). "Animal Ecology." Methuen, London.

Emerson, A. E. (1956). Regenerative behavior and social homeostasis of termites. *Ecology* **27**, 248–258.

Ettema, C. H., and Bongers, T. (1993). Characterization of nematode colonization and succession in disturbed soil using the Maturity Index. *Biol. Fertil. Soils* **16**, 79–85.

Farrah, S. R., and Bitton, G. (1990). Viruses in the soil environment. *In* "Soil Biochemistry 6" (J.-M. Bollag and G. Stotzky, eds.), pp. 529–556. Dekker, New York.

Fender, W. M. (1995). Native earthworms of the Pacific Northwest: An ecological overview. *In* "Ecology and Biogeography of Earthworms in North America" (P. F. Hendrix, ed.), pp. 53–66. Lewis, Boca Raton, Florida.

Fenster, C. R., and Peterson, G. A. (1979). Effects of no-tillage fallow as compared to conventional tillage in a wheat-fallow system. Research Bulletin 289. Agricultural Experiment Station, University of Nebraska, Lincoln.

Ferguson, L. M. (1990). Insecta: Diplura. *In* "Soil Biology Guide" (D. L. Dindal, ed.), pp. 951–964. Wiley, New York.

Fitter, A. H. (1985). Functional significance of root morphology and root system architecture. *In* "Ecological Interactions in Soil; Plants, Microbes and Animals" (A. H. Fitter, D. Atkinson, D. J. Read, and M. B. Usher, eds.), pp. 87–106. Blackwell, Oxford.

Fitter, A. H. (1991). The ecological significance of root system architecture: An economic approach. *In* "Plant Root Growth: An Ecological Perspective" (D. Atkinson, ed.), pp. 229–243. Blackwell, Oxford.

FitzPatrick, E. A. (1984). "Micromorphology of Soils." Chapman & Hall, London.

Fogel, R. (1985). Roots as primary producers in below-ground ecosystems. *In* "Ecological Interactions in Soil: Plants, Microbes and Animals" (A. H. Fitter, D. Atkinson, D. J. Read, and M. B. Usher, eds.), pp. 23–36. Blackwell, Oxford.

Fogel, R. (1991). Root system demography and production in forest ecosystems. *In* "Plant Root Growth: An Ecological Perspective" (D. Atkinson, ed.), pp. 89–101. Blackwell, Oxford.

Fogel, R., and Lussenhop, J. (1991). The University of Michigan soil biotron: A platform for soil biology research in a natural forest. *In* "Plant Root Growth: An Ecological Perspective" (D. Atkinson, ed.), pp. 61–73. Blackwell, Oxford.

Foissner, W. (1987). Soil protozoa: Fundamental problems, ecological significance, adaptations in ciliates and testaceans, bioindicators, and guide to the literature. *Prog. Protistol.* **2**, 69–212.

Foissner, W. (1994). Soil protozoa as bioindicators in ecosystems under human influence. *In* "Soil Protozoa" (J. F. Darbyshire, ed.), pp. 147–193. CAB International, Wallingford, U.K.

Foster, R. C. (1985). *In situ* localization of organic matter in soils. *Quaestiones Entomologicae* **21**, 609–633.

Foster, R. C. (1988). Microenvironments of soil microorganisms. *Biol. Fertil. Soils* **6**, 189–203.

Foster, R. C. (1994). Microorganisms and soil aggregates. *In* "Soil Biota" (C. E. Pankhurst, B. M. Doube, V. V. S. R. Gupta, and P. R. Grace, eds.), pp. 144–155. CSIRO, Melbourne, Australia.

Foster, R. C., and Dormaar, J. F. (1991). Bacteria-grazing amoebae *in situ* in the rhizosphere. *Biol. Fertil. Soils* **11**, 83–87.

Foster, R. C., Rovira, A. D., and Cock, T. W. (1983). "Ultrastructure of the Root Soil Interface." American Phytopathology Society, St. Paul, Minnesota.

Fragoso, C., James, S. W., and Borges, S. (1995). Native earthworms of the north neotropical region: Current status and controversies. *In* "Ecology and

Biogeography of Earthworms in North America" (P. F. Hendrix, ed.), pp. 67–103. Lewis, Boca Raton, Florida.

Francé, R. H. (1921). "Das Edaphon." Arbeiten an der Biologisches Institut der Muenchen, Stuttgart.

Freckman, D. W., ed. (1994). Life in the soil/soil biodiversity: Its importance to ecosystem processes. Report on a workshop held at The Natural History Museum, London. NREL, Colorado State University, Ft. Collins, Colorado.

Freckman, D. W., and Ettema, C. H. (1993). Assessing nematode communities in agroecosystems of varying human intervention. *Agric. Ecosyst. Environ.* **45**, 239–261.

Freckman, D. W., and Virginia, R. A. (1989). Plant-feeding nematodes in deep-rooting desert ecosystems. *Ecology* **70**, 1665–1678.

Frost, S. W. (1942). "General Entomology." McGraw-Hill, New York.

Gadgil, R. L., and Gadgil, P. D. (1975). Suppression of litter decomposition by mycorrhizal roots of *Pinus radiata*. *N. Z. J. For. Sci.* **5**, 33–41.

Gill, H. S., and Abrol, I. P. (1986). Salt affected soils and their amelioration through afforestation. *In* "Amelioration of Soil by Trees" (R. T. Prinsley and M. J. Swift, eds.), pp. 43–53. Commonwealth Science Council. Marlborough House, London.

Gilmore, S. K. (1972). Collembola predation on nematodes. *Search Agric.* **1**, 1–12.

Gist, C. S., and Crossley, D. A., Jr. (1973). A method for quantifying pitfall trapping. *Environ. Entomol.* **2**, 951–952.

Górny, M., and Grüm, L. (1993). "Methods in Soil Zoology." PWN-Polish Scientific, Warszawa, Poland.

Griffiths, B. S. (1994). Soil nutrient flow. *In* "Soil Protozoa" (J. F. Darbyshire, ed.), pp. 65-91. CAB International, Wallingford, U.K.

Griffiths, B. S., and Caul, S. (1993). Migration of bacterial-feeding nematodes, but not protozoa, to decomposing grass residues. *Biol. Fertil. Soils* **15**, 201–207.

Griffiths, E. (1965). Micro-organisms and soil structure. *Biol. Rev.* **40**, 129–142.

Grodzinski, W., and Yorks, T. P. (1981). Species and ecosystem-level bioindicators of airborne pollution: An analysis of two major studies. *Water Air Soil Pollut.* **16**, 33–53.

Guillet, B. (1990). Le vieillissement des matières organiques et des associations organominérales des andosols et des podsols. *Sci. Sol* **28**, 285–299.

Gupta, V. V. S. R. (1989). Microbial biomass sulfur and biochemical mineralization of sulfur in soils. Ph.D. Thesis. University of Saskatchewan, Saskatoon.

Gupta, V. V. S. R., and Germida, J. J. (1988). Populations of predatory protozoa in field soils after 5 years of elemental S fertilizer application. *Soil Biol. Biochem.* **20**, 787–791.

Gupta, V. V. S. R., and Germida, J. J. (1989). Influence of bacterial–amoebal interactions on sulfur transformations in soil. *Soil Biol. Biochem.* **21**, 921–930.

Haas, H. J., Evans, C. E., and Miles, E. F. (1957). Nitrogen and carbon changes in Great Plains soils as influenced by cropping and soil treatments. Technical Bulletin No. 1164, USDA, Washington, D.C.

Hadas, A. (1979). Heat capacity *In* "The Encyclopedia of Soil Science, Part 1: Physics, Chemistry, Biology, Fertility, and Technology" (R. W. Fairbridge and C. W. Finkl, Jr. eds), p. 189. Dowden, Hutchinson & Ross, Inc., Stroudsburg, Pennsylvania.

Haimi, J., and Einbork, M. (1992). Effects of endogeic earthworms on soil processes and plant growth in coniferous forest soil. *Biol. Fertil. Soils* **13**, 6–10.

Haines, B. L., and Swank, W. T. (1988). Acid precipitation effects on forest processes. *In* "Forest Hydrology and Ecology at Coweeta" (W. T. Swank and D. A. Crossley, Jr., eds.), pp. 359–366. Springer-Verlag, New York.

Hairston, N. G., Jr., and Hairston, N. G., Sr. (1993). Cause–effect relationships in energy flow, trophic structure, and interspecific interactions. *Am. Nat.* **142**, 379–411.

Hall, S. J., and Raffaelli, D. G. (1993). Food webs: Theory and Reality. *Adv. Ecol. Res.* **24**, 187–239.

Hallsworth, E. G., and Crawford, D. V. (1965). "Experimental Pedology." Butterworth, London.

Hansson, A., Andrén, O., and Steen, E. (1991). Root production of four arable crops in Sweden and its effect on abundance of soil organisms. *In* "Plant Root Growth: An Ecological Perspective" (D. Atkinson, ed.), pp. 247–266. Blackwell, Oxford.

Hanula, J. L. (1995). Relationship of wood-feeding insects and coarse woody debris. *In* "Biodiversity and Coarse Woody Debris in Southern Forests" (J. W. McMinn and D. A. Crossley, Jr., eds.), in press. U.S.D.A. Forest Service, Asheville, North Carolina.

Harris, R. F., Chesters, G., Allen, O. N., and Attoe, O. J. (1964). Mechanisms involved in soil aggregate stabilization by fungi and bacteria. *Soil Sci. Soc. Am. Proc.* **28**, 529–532.

Harris, W. F., Kinerson, R. S., Jr., and Edwards, N. T. (1977). Comparison of below-ground biomass of natural deciduous forest and loblolly pine plantations. *Pedobiologia* **7**, 369–381.

Harvey, R. W., Kinner, N. E., Bunn, A., MacDonald, D., and Metge, D. (1995). Transport behavior of groundwater protozoa and protozoan-sized micro-spheres in sandy aquifer sediments. *Appl. Environ. Microbiol.* **61**, 209–271.

Hattori, T. (1994). Soil micro environment. *In* "Soil Protozoa" (J. F. Darbyshire, ed.), pp. 43–64. CAB International, Wallingford, U.K.

Hawksworth, D. L. (1991a). "The Biodiversity of Microorganisms and Invertebrates." C.A.B. International, Wallingford, U.K.

Hawksworth, D. L. (1991b). The fungal dimension of biodiversity: Magnitude, significance and conservation. *Mycol. Res* **95**, 641–655.

Heal, O. W., and Dighton, J. (1985). Resource quality and trophic structure in the soil system. *In* "Ecological Interactions in Soil: Plants, Microbes and Animals" (A. H. Fitter, D. Atkinson, D. J. Read, and M. B. Usher, eds.), pp. 339–354. Blackwell, Oxford.

Hedley, M. J., and Stewart, J. W. B. (1982). Method to measure microbial phosphate in soil. *Soil Biol. Biochem.* **14**, 377-385.

Helal, H. M., and Sauerbeck, D. (1991). Short term determination of the actual respiration rate of intact plant roots. *In* "Plant Roots and Their Environment" (B. L. Michael and H. Persson, eds.), pp. 88–92. Elsevier, Amsterdam.

Henderson, L. J. (1913). "The Fitness of the Environment." Beacon Hill Press, Boston.

Henderson, L. S. (1952). Household insects. *In* "Insects, The Yearbook of Agriculture 1952" (Stefferud. A. ed.), pp. 469–475. United States Government Printing Office, Washington, D.C.

Hendrick, R. L. and Pregitzer, K. S. (1992). The demography of fine roots in a Northern hardwood forest. *Ecology* **73**, 1094–1104.

Hendrix, P. F. (1995). "Earthworm Ecology and Biogeography in North America." Lewis, Boca Raton, Florida.

Hendrix, P. F., Parmelee, R. W., Crossley, D. A., Jr., Coleman, D. C., Odum, E. P., and Groffman, P. (1986). Detritus food webs in conventional and no-tillage agroecosystems. *Bioscience* **36**, 374–380.

Hendrix, P. F., Crossley, D. A., Jr., Coleman, D. C. Parmelee, R. W., and Beare, M. H. (1987). Carbon dynamics in soil microbes and fauna in conventional and no-tillage agroecosystems. *INTECOL Bull.* **15**, 59–63.

Hendrix, P. F., Crossley, D. A, Jr., Blair, J. M., and Coleman, D. C. (1990). Soil biota as components of sustainable agroecosystems. *In* "Sustainable Agricultural Systems" (C. A. Edwards, R. Lal, P. Madden, R. H. Miller, and G. House, eds.), pp. 637–654. Soil and Water Conservation Society, Ankeny, Iowa.

Hendrix, P. F., Coleman, D. C., and Crossley, D. A., Jr. (1992). Using knowledge of soil nutrient cycling processes to design sustainable agriculture. *J. Sust. Agric.* **2**, 63–82.

Hillel, D. J. (1991). "Out of the Earth: Civilization and the Life of the Soil." Free Press, New York.

Hiltner, L. (1904). Über neuere erfahrungen und probleme auf dem gebiet der boden-bakteriologie und unter besonderer berücksichtigung der gründüngung und brache. *Arb. DLG* **98**, 59–78.

Holland, E. A., and Coleman, D. C. (1987). Litter placement effects on microbial and organic matter dynamics in an agroecosystem. *Ecology* **68**, 425–433.

Hölldobler, B., and Wilson, E. O. (1990). "The Ants." Belknap Press, Harvard University, Cambridge.

Houghton, R. A. (1990). The global effects of deforestation. *Environ. Sci. Technol.* **24**, 414–422.

Houghton, R. A., Hobbie, J. E., Melillo, J. M., Moore, B., Peterson, B. J., Shaver, G. R., and Woodwell, G. M. (1983). Changes in the carbon content of terrestrial biota and soils between 1860 and 1980: A net release of CO_2 to the atmosphere. *Ecol. Monogr.* **53**, 235–262.

Houghton, R. A., Boone, R. D., Fruci, J. R., Hobbie, J. E., Melillo, J. M., and Palm, C. A. (1987). The flux of carbon from terrestrial ecosystems to the atmosphere in 1980 due to changes in land use: Geographic distribution of the global flux. *Tellus* **39B**, 122–139.

House, G. J., Stinner, B. R., and Crossley, D. A., Jr. (1984). Nitrogen cycling in conventional and no-tillage agroecosystems: Analysis of pathways and processes. *J. Appl. Ecol.* **21**, 991–1012.

Huhta, V., Wright, D.H., and Coleman, D.C. (1989). Characteristics of defaunated soil. I. A comparison of three techniques applied to two different forest soils. *Pedobiologia* **33**, 417–426.

Humphreys, W. F. (1979). Production and respiration in animal populations. *J. Anim. Ecol.* **48**, 427–453.

Hunt, H. W., Coleman, D. C., Ingham, E. R., Ingham, R. E., Elliott, E. T., Moore, J. C., Rose, S. L., Reid, C. P. P., and Morley, C. R. (1987). The detrital food web in a shortgrass prairie. *Biol. Fertil. Soils* **3**, 57–68.

Hunter, P. E., and Rosario, R. M. T. (1988). Associations of Mesostigmata with other arthropods. *Annu. Rev. Entomol.* **33**, 393–413.

Ingham, E. R., and Horton, K. A. (1987). Bacterial, fungal and protozoan responses to chloroform fumigation in stored soil. *Soil Biol. Biochem.* **19**, 545–550.

Ingham, E. R., and Klein, D. A. (1982). Relationship between fluorescein diacetate-stained hyphae and oxygen utilization, glucose utilization, and biomass of submerged fungal batch cultures. *Appl. Environ. Microbiol.* **44**, 363–370.

Ingham, E. R., Trofymow, J. A., Ames, R. N., Hunt, H. W., Morley, C. R., Moore, J. C., and Coleman, D. C. (1986a). Trophic interactions and nitrogen cycling in a semiarid grassland soil. Part I. Seasonal dynamics of the natural populations, their interactions and effects on nitrogen cycling. *J. Appl. Ecol.* **23**, 597–614.

Ingham, E. R., Trofymow, J. A., Ames, R. N., Hunt, H. W., Morley, C. R., Moore, J. C., and Coleman, D. C. (1986b). Trophic interactions and nitrogen cycling in a semiarid grassland soil II. System responses to removal of different groups of soil microbes or fauna. *J. Appl. Ecol.* **23**, 615–630.

Ingham, E. R., Coleman, D. C., and Moore, J. C. (1989). An analysis of food web structure and function in a shortgrass prairie, mountain meadow and lodgepole pine forest. *Biol. Fertil. Soils* **8**, 29–37.

Ingham, R. E., Trofymow, J. A., Ingham, E. R., and Coleman, D. C. (1985). Interactions of bacteria, fungi, and their nematode grazers: Effects on nutrient cycling and plant growth. *Ecol. Monogr.* **55**, 119–140.

Insam, H. (1990). Are the soil microbial biomass and basal respiration governed by the climatic regime? *Soil Biol. Biochem.* **22**, 525–532.

Insam, H. and Domsch, K. H. (1988). Relationship between soil organic carbon and microbial biomass on chronosequences of reclamation sites. *Microb. Ecol.* **15**, 177–188.

Jacot, A. P. (1936). Soil structure and soil biology. *Ecology* **17**, 359–379.

James, S. W. (1995). Systematic, biogeography and ecology of nearctic earthworms from eastern, central, southern and southwestern United States. *In* "Ecology and Biogeography of Earthworms in North America" (P. F. Hendrix, ed.), pp. 29–52. Lewis, Boca Raton, Florida.

Jastrow, J. D., and Miller, R. M. (1991). Methods for assessing the effects of biota on soil structure. *Agric. Ecosyst. Environ.* **34**, 279–303.

Jenkinson, D. S. (1966). Studies on the decomposition of plant material in soil, II. Partial sterilization of soil and the soil biomass. *J. Soil Sci.* **17**, 280–302.

Jenkinson, D. S. (1988). Determination of microbial biomass carbon and nitrogen in soil. *In* "Advances in Nitrogen Cycling in Agricultural Ecosystems" (J. R. Wilson, ed.). CAB International, Wallingford, U.K.

Jenkinson, D. S., and Parry, L. C. (1989). The nitrogen cycle in the Broadbalk wheat experiment: A model for the turnover of nitrogen through the soil microbial biomass. *Soil Biol. Biochem.* **21**, 535–541.

Jenkinson, D. S., and Powlson, D. S. (1976). The effects of biocidal treatments on metabolism in soil. V. A method for measuring soil biomass. *Soil Biol. Biochem.* **8**, 209–213.

Jenkinson, D. S., Powlson, D. S., and Wedderburn, F. W. M. (1976). The effects of biocidal treatments on metabolism in soil. III. The relationship between soil biovolume, measured by optical microscopy, and the flush of decomposition caused by fumigation. *Soil Biol. Biochem.* **8**, 189–202.

Jenkinson, D. S., Adams, D. E., and Wild, A. (1991). Model estimates of CO_2 emissions from soil in response to global warming. *Nature (London)* **351**, 304–306.

Jenny, H. (1941). "Factors of Soil Formation." McGraw-Hill, New York.

Jenny, H. (1980). "The Soil Resource: Origin and Behavior." Ecological Studies 37. Springer-Verlag, New York.

Jenny, H., and Grossenbacher, K. (1963). Root-soil boundary zones as seen in the electron microscope. *Soil Sci. Soc. Am. Proc.* **27**, 273–277.

Joergensen, R. G., Anderson, T.-H., and Wolters, V. (1995). Carbon and nitrogen relationships in the microbial biomass of soils in beech (*Fagus sylvatica* L.) forests. *Biol. Fertil. Soils* **19**, 141–147.

Johnson, D. L. (1990). Biomantle evolution and the redistribution of earth materials and artifacts. *Soil Sci.* **149**, 84–102.

Jones, P. C. T., and Mollison, J. E. (1948). A technique for the quantitative estimation of soil microorganisms. *J. Gen. Microbiol.* **2**, 54–69.

Jongerius, A., ed. (1964). "Soil Micromorphology." Elsevier, Amsterdam.

Kaczmarek, M. (1993). Apparatus and tools for the extraction of animals from the soil. *In* "Methods in Soil Zoology" (M. Górny and L. Grüm, eds.), pp. 112–141. PWN— Polish Scientific Publishers. Warszawa, Poland.

Kanoko, N. (1988). Feeding habits and cheliceral size of oribatid mites in cool temperate forest soils in Japan. *Rev. Ecol. Biol. Sol* **25**, 353–363.

Karg, W. (1982). Investigations of habitat requirements, geographical distributions and origin of predatory mite genera in the Cohort Gamasina for use as bioindicators. *Pedobiologia* **24**, 241–247.

Kevan, D. K. M., and Scudder, G. G. E. (1989). "Illustrated keys to the families of terrestrial arthropods of Canada 1. Myriapods (Millipedes, Centipedes, etc.)" Biological Survey of Canada (Terrestrial Arthropods), Ottawa.

Kilbertus, G. (1980). Etudes des microhabitats contenus dans les agrégats du sol leur relation avec la biomasse bactérienne et la taille des procaryotes présents. *Rev. Ecol. Biol. Sol* **17**, 543–557.

Kilbertus, G. and Vannier, G. (1981). Relations microflore-microfaune dans la grotte de Sainte-Catherine (*Pyrénéés ariegeoises*). II. Le regime alimentaire de *Tomocerus minor* (Lubbock) et *Tomocerus problematicus* Cassagnau (Insectes Collemboles). *Rev. Ecol. Biol. Sol* **18**, 319–338.

Kimble, J. M., and Levine, E. R. (1994). The Nairobi Conference: Topics, results, and research needs. *Trans. 15th World Congr. Soil Sci, INEGI CNA,* 151–162.

Kjøller, A., and Struwe, S. (1994). Analysis of fungal communities on decomposing beech litter. *In* "Beyond the Biomass" (K. Ritz, J. Dighton, and K. E. Giller, eds.), pp. 191–200. Wiley, Chichester.

Kleiber, M. (1961). "The Fire of Life." Wiley, New York.

Kozlowska, J., and Wasilewska, L. (1993). Nematoda. *In* "Methods in Soil Zoology" (M. Górny and L. Grüm, eds.) pp. 163–183. PWN—Polish Scientific Publishers, Warszawa, Poland.

Krantz, G. W. (1978). "A Manual of Acarology," 2nd Ed. Oregon State University Book Stores, Corvallis, Oregon.

Kubiëna, W. L. (1938). "Micropedology." Collegiate Press, Ames, Iowa.

Kucey, R. M. N., and Paul, E. A. (1982). Carbon flow, photosynthesis, and N_2 fixation in mycorrhizal and nodulated faba beans (*Vicia faba* L.). *Soil Biol. Biochem.* **14**, 407–412.

Kühnelt, W. (1958). Zoogenic crumb-formation in undisturbed soils (in German). *Sonderdruck aus Tagungsberichte* **13**, 193–199.

Kühnelt, W. (1976). "Soil Biology." Michigan State University, East Lansing, Michigan.

Kuikman, P., and van Veen, J. A. (1989). The impact of protozoa on the availability of bacterial nitrogen to plants. *Biol. Fertil. Soils* **8**, 13–18.

Kuikman, P. J., Jansen, A. G., van Veen, J. A., and Zehnder, A. J. B. (1990). Protozoan predation and the turnover of soil organic carbon and nitrogen in the presence of plants. *Biol. Fertil. Soils* **10**, 22–28.

Kurcheva, G. F. (1960). Role of soil organisms in the breakdown of oak litter. *Pochvovedeniye* **4**, 16–23 (in Russian).

Kurcheva, G. F. (1964). Wirbellose Tiere als Faktor der Zersetzung von Waldstreu. *Pedobiologia* **4**, 7–30.

Lal, R., Kimble, J., Levine, E., and Stewart, B. A., eds. (1995) "Soils and Global Change." CRC Lewis Publishers, Boca Raton, Florida.

Lartey, R. T., Curl, E. A., and Peterson, C. M. (1994). Interactions of mycophagous collembola and biological control fungi in the suppression of *Rhizoctonia solani*. *Soil Biol. Biochem.* **26**, 81–88.

Lashof, D. A. (1989). The dynamic greenhouse: Feedback processes that may influence future of atmospheric trace gases and climate change. *Clim. Change* **14**, 213–242.

Lavelle, P., and Martin, A. (1992). Small-scale and large-scale effects of endogeic earthworms on soil organic matter dynamics in soils of the humid tropics. *Soil Biol. Biochem.* **24**, 1491–1498.

Lavelle, P., Blanchart, E., and Martin, A. (1992). Impact of soil fauna on the properties of soils in the humid tropics. *In* "Myths and Science of Soils of the Tropics" (R. Lal and P. Sanchez, eds.), pp. 157–185. Soil Science Society of America, Madison, Wisconsin.

Lee, K. E. (1985). "Earthworms: Their Ecology and Relationships with Soils and Land Use." Academic Press, Sydney, Australia.

Lee, K. E., and Foster, R. C. (1991). Soil fauna and soil structure. *Aust. J. Soil Res.* **29**, 745–775.

Lee, K. E., and Pankhurst, C. E. (1992). Soil organisms and sustainable productivity. *Aust. J. Soil Res.* **30**, 855–892.

Lee, K. E., and Wood, T. G. (1971). "Termites and Soils." Academic Press, London and New York.

Leetham, J. W., McNary, T. J., Dodd, J. L., and Lauenroth, W. K. (1982). Response of soil nematodes, rotifers, and tardigrades to three levels of season-long sulphur dioxide exposure. *Water Air Soil Pollut.* **18**, 343–356.

Levine, N. D., Corliss, J. O., Cox, F. E. G., Deroux, D., Grain, J., Honigberg, B. M., Leedale, G. F., Loeblich, R., III, Lom, J., Lynn, D., Merinfeld, E. G., Page, F. C., Poljansky, G., Spraque, V., Vavra, J., Wallace, F. G., and Wieser, J. (1980). A new revised classification of the protozoa. *J. Protozool.* **27**, 36–58.

Lindeman, R. L. (1942). The trophic-dynamic aspect of ecology. *Ecology* **23**, 399–418.

Linden, D. R., Hendrix, P. F., Coleman, D. C., and van Vliet, P. C. J. (1994). Faunal indicators of soil quality. *In* "Defining Soil Quality for a Sustainable Environment." (J. W. Doran, D. C. Coleman, D. F. Bezdicek, and B. A. Stewart, eds.), SSSA Special Publication, pp. 91–106. American Society of Agronomy, Madison, Wisconsin.

Lotka, A. J. (1925). "Elements of Physical Biology." Williams & Wilkins, Baltimore, Maryland.

Lousier, J. D., and Bamforth, S. S. (1990). Soil protozoa. *In* "Soil Biology Guide" (D. L. Dindal, ed.), pp. 97–136. Wiley, New York.

Lousier, J. D., and Parkinson, D. (1981). Evaluation of a membrane filter technique to count soil and litter Testacea. *Soil Biol. Biochem.* **13**, 209–213.

Lousier, J. D., and Parkinson, J. (1984). Annual population dynamics and production ecology of Testacea (Protozoa, Rhizopoda) in an aspen woodland soil. *Soil Biol. Biochem.* **16**, 103–114.

Lovelock, J. E. (1979). "Gaia: A New Look at Life on Earth." Oxford Univ. Press, Oxford and London.

Lovelock, J. E. (1988). "The Ages of Gaia." Norton, New York.

Lussenhop, J. (1976). Soil arthropod response to prairie burning. *Ecology* **57**, 88–98.

Lussenhop, J. (1992). Mechanisms of microarthropod–microbial interactions in soil. *Adv. Ecol. Res.* **23**, 1–33.

Luxton, M. (1967). The ecology of saltmarsh Acarina. *J. Anim. Ecol.* **36**, 257–277.

Luxton, M. (1972). Studies on the oribatid mites of a Danish beech forest. I. Nutritional biology. *Pedobiologia* **12**, 434–463.

Luxton, M. (1975). Studies on the oribatid mites of a Danish beech forest. II. Biomass, calorimetry and respirometry. *Pedobiologia* **15**, 161–200.

Luxton, M. (1979). Food and energy processing by oribatid mites. *Rev Écol. Biol. Sol* **16**, 103–111.

Luxton, M. (1981a). Studies on the astigmatic mites of a Danish beech wood soil. *Pedobiologia* **22**, 29–38.

Luxton, M. (1981b). Studies on the prostigmatic mites of a Danish beech wood soil. *Pedobiologia* **22**, 277–303.

Lynch, J. M. (1990). "The Rhizosphere." Wiley (Interscience), Chichester.

MacArthur, R. H. (1972). "Geographical Ecology: Patterns in the Distribution of Species." Harper & Row, New York.

McBrayer, J. F. (1973). Exploitation of deciduous leaf litter by *Apheloria montana*. *Pedobiologia* **13**, 90–98.

Macfadyen, A. (1969). The systematic study of soil ecosystems. *In* "The Soil Ecosystem" (J. G. Sheals, ed.), pp. 191-197. The Systematics Association, London.

McGill, W. B., and Cole, C. V. (1981). Comparative aspects of cycling of organic C, N, S, and P through soil organic matter. *Geoderma* **26**, 267–286.

Mackie-Dawson, L. A., and Atkinson, D. (1991). Methodology for the study of roots in field experiments and the interpretation of results. *In* "Plant Root Growth: An Ecological Perspective" (D. Atkinson, ed.), pp. 25–47. Blackwell, London.

Malloch, D. W., Pirozynski, K. A., and Raven, P. A. (1980). Ecological and evolutionary significance of mycorrhizal symbioses in vascular plants (a review). *Proc. Natl. Acad. Sci. U.S.A.* **77**, 2113–2118.

Mann, L. K. (1986). Changes in soil organic carbon storage after cultivation. *Soil Sci.* **142**, 279–288.

Marinissen, J. C. Y., and Dexter, A. R. (1990). Mechanisms of stabilization of earthworm casts and artificial casts. *Biol. Fertil. Soils* **9**, 163–167.

Marshall, V. G. (1977). Effects of manures and fertilizers on soil fauna: A review. Special Publication 3. Commonwealth Bureau of Soils, Slough, U.K.

Marshall, V. G., Reeves, R. M., and Norton, R. A. (1987). Catalogue of the Oribatida (Acari) of the continental United States and Canada. *Memoirs of the Entomol. Soc. Canada* **139**, 418.

Martin, J. K., and Kemp, J. R. (1986). The measurement of carbon transfers within the rhizosphere of wheat grown in field plots. *Soil Biol. Biochem.* **18**, 103–107.

Martin, M. M. (1984). The role of ingested enzymes in the digestive processes of insects. *In* "Invertebrate–Microbial Interactions" (J. M. Anderson, A. D. M. Rayner, and D. W. H. Walton, eds.), pp. 155–172. Cambridge Univ. Press, Cambridge.

Maynard, D. G., Stewart, J. W. B., and Bettany, J. R. (1984). Sulfur cycling in grassland and parkland soils. *Biogeochemistry* **1**, 97–111.

Meentemeyer, V. (1978). Macroclimate and lignin control of decomposition. *Ecology* **59**, 465–472.

Melillo, J. M., Aber, J. D., and Muratore, J. F. (1982). Nitrogen and lignin control of hardwood leaf litter decomposition dynamics. *Ecology* **63**, 621–626.

Metcalf, C. L., and Flint, W. P. (1939). "Destructive and Useful Insects." 2nd Ed. McGraw-Hill, New York.

Michener, C. D., and Michener, M. H. (1951). "American Social Insects." Van Nostrand, New York.

Milchunas, D. G., Lauenroth, W. K., Singh, J. S., Cole, C. V., and Hunt, H. W. (1985). Root turnover and production by ^{14}C dilution: implications of carbon partitioning in plants. *Plant Soil* **88**, 353–365.

Mitchell, M. J., and Parkinson, D. (1976). Fungal feeding of oribatid mites (Acari: cryptostigmata) in an aspen woodland soil. *Ecology* **57**, 302–312.

Monz, C. A., Reuss, D. E., and Elliott, E. T. (1991). Soil microbial biomass carbon and nitrogen estimates using 2450 MHz microwave irradiation or chloroform fumigation followed by direct extraction. *Agric. Ecosyst. Environ.* **34**, 55–63.

Moore, J. C., and de Ruiter, P. C. (1991). Temporal and spatial heterogeneity of trophic interactions within below-ground food webs. *Agric. Ecosyst. Environ.* **34**, 371–397.

Moore, J. C., St. John, T. V., and Coleman, D. C. (1985). Ingestion of vesicular arbuscular mycorrhizal hyphae and spores by soil microarthropods. *Ecology* **66**, 1979–1981.

Moore, J. C., Ingham, E. R., and Coleman, D. C. (1987). Inter- and intraspecific feeding selectivity of *Folsomia candida* (Willem) (Collembola, Isotomidae) on fungi. *Biol. Fertil. Soils* **5**, 6–12.

Moore, J. C., Walter, D. E., and Hunt, H. W. (1988). Arthropod regulation of micro- and mesobiota in belowground food webs. *Annu. Rev. Entomol.* **33**, 419–439.

Mosier, A., Schimel, D., Valentine, D., Bronson, K., and Parton, W. (1991). Methane and nitrous oxide fluxes in native, fertilized and cultivated grasslands. *Nature (London)* **350**, 330–332.

Mueller, B. R., Beare, M. H., and Crossley, D. A., Jr. (1990). Soil mites in detrital food webs of conventional and no-tillage agroecosystems. *Pedobiologia* **34**, 389–401.

Müller, P. E. (1887). "Studien über die natürlichen Humusformen und deren Einwirkungaut Vegetation und Boden" Springer, Berlin.

Nadelhoffer, K. J., Aber, J. D., and Melillo, J. M. (1985). Fine roots, net primary production, and soil nitrogen availability: A new hypothesis. *Ecology* **66**, 1377–1390.

Nannipieri, P. (1994). The potential use of soil enzymes as indicators of productivity, sustainability and pollution. *In* "Soil Biota: Management in Sustainable Farming" (C. E. Pankhurst, B. M. Doube, V. V. S. R. Gupta, and P. R. Grace, eds.), pp. 238–244. CSIRO, Melbourne, Australia.

Nannipieri, P., Grego, S., and Ceccanti, B. (1990). Ecological significance of the biological activity in soil. *In* "Soil Biochemistry, Volume 6." (J.-M. Bollag and G. Stotzky, eds.), pp. 293–356. Dekker, New York.

Neher, D. A., Peck, S. L., Rawlings, J. O., and Campbell, C. L. (1995). Variability of plant parasitic and free-living nematode communities within and between fields and geographic regions of North Carolina. *Plant Soil* **170**, 167–181.

Nelson, D. R., and Higgins, R. P. (1990). Tardigrada. *In* "Soil Biology Guide" (D. L. Dindal, ed.), pp. 393–419. Wiley, New York.

Neuhauser, E. F., and Hartenstein, R. (1978). Phenolic content and palatability of leaves and wood to soil isopods and diplopods. *Pedobiologia* **18**, 99–109.

Newell, K. (1984a). Interaction between two decomposer basidiomycetes and a collembolan under Sitka spruce: Distribution, abundance and selective grazing. *Soil Biol. Biochem.* **16**, 227–233.

Newell, K. (1984b). Interaction between two basidiomycetes and Collembola under Sitka spruce: grazing and its potential effects on fungal distribution and litter decomposition. *Soil Biol. Biochem.* **16**, 235–240.

Newell, S. Y., and Fallon, R. D. (1991). Toward a method for measuring instantaneous fungal growth rates in field samples. *Ecology* **72**, 1547–1559.

Newman, A. S., and Norman, A. G. (1943). An examination of thermal methods for following microbiological activity in soil. *Soil Sci. Soc. Am. Proc.* **8**, 250–253.

Newman, R. H., and Tate, K. R. (1980). Soil phosphorus characterization by ^{31}P nuclear magnetic resonance. *Commun. Soil Sci. Plant Anal.* **11**, 835–842.

Nielson, G. A., and Hole, F. D. (1964). Earthworms and the development of coprogenous A1-horizons in forest soils of Wisconsin. *Soil Sci. Soc. Am. Proc.* **28**, 426–430.

Norton, R. A., and Behan-Pelletier, V. M. (1991). Calcium carbonate and calcium oxalate as cuticular hardening agents in oribatid mites (Acari: Oribatida). *Can. J. Zool.* **69**, 1504–1511.

Norton, R. W., Bonamo, P. M., Grierson, J. D., and Shear, W. A. (1987). Fossil mites from the Devonian of New York State. *In* "Progress in Acarology" (G. P. Channabasavanna and C. A. Viraktamath, eds.), Vol. 1, pp. 271–277. Oxford & IBH Publ., New Delhi.

Oades, J. M. (1984). Soil organic matter and structural stability: mechanisms and implications for management. *Plant Soil* **76**, 319–337.

Oades, J. M., and Waters, A. G. (1991). Aggregate hierarchy in soils. *Aust. J. Soil Res.* **29**, 815–828.

Oades, J. M., Gillman, G. P., Uehara, G., Hue, N. V., van Noordwijk, M., Robertson, G. P., and Wada, K. (1989). Interactions of soil organic matter and variable-charge clays. *In* "Dynamics of Soil Organic Matter in Tropical Ecosystems" (D. C. Coleman, J. M. Oades, and G. Uehara, eds.), pp. 69–95. Univ. of Hawaii Press, Honolulu.

Odum, E. P. (1971). "Fundamentals of Ecology." 3rd Ed. Saunders, Philadelphia, Pennsylvania.

Odum, E. P., and Biever, L. J. (1984). Resource quality, mutualism and energy partitioning in food chains. *Am. Nat.* **124**, 360–376.

Olson, J. S. (1963). Energy storage and the balance of producers and decomposers in ecological systems. *Ecology* **44**, 322–331.

Olson, J. S., and Crossley, D. A., Jr. (1963). Tracer studies of the breakdown of forest litter. *In* "Radioecology" (V. Schultz and A.W. Klement Jr., eds,), pp. 411–416. Reinhold Publishing Corporation, New York.

Olson, J. S., Watts, J. A., and Allison, L. J. (1983). "Carbon in Live Vegetation of the Major World Ecosystems." National Technical Information Service, Springfield, Virginia.

O'Neill, R. V., DeAngelis, D. L., Waide, J. B., and Allen, T. F. H. (1986). "A Hierarchical Concept of Ecosystems." Princeton Univ. Press, Princeton, New Jersey.

Pantastico-Caldas, M., Duncan, K. E., Istock, C. A., and Bell, J. A. (1992). Population dynamics of bacteriophage and *Bacillus subtilis* in soil. *Ecology* **73**, 1888–1902.

Parker, L. W., Santos, P. F., Phillips, J., and Whitford, W. G. (1984). Carbon and nitrogen dynamics during the decomposition of litter and roots of a Chihuahuan desert annual *Lepidium lasiocarpum*. *Ecological Monographs* **54**, 339–360.

Parkinson, D., and Coleman, D. C. (1991). Microbial populations, activity and biomass. *Agric. Ecosyst. Environ.* **34**, 3–33.

Parkinson, D., Gray, T. R. G., and Williams, S. T. (1971). "Methods for Studying the Ecology of Soil Microorganisms. IBP Handbook Number 19." Blackwell, Oxford.

Parmelee, R. W., Beare, M. H., Cheng, W., Hendrix, P. F., Rider, S. J., Crossley, D.A., Jr., and Coleman, D. C. (1990). Earthworms and enchytraeids in conventional and no-tillage agroecosystems: A biocide approach to assess their role in organic matter breakdown. *Biol. Fertil. Soils* **10**, 1–10.

Parton, W. J., Schimel, D. S., Cole, C. V., and Ojima, D. S. (1987). Analysis of factors controlling soil organic matter levels in Great Plains grasslands. *Soil Sci. Soc. Am. J.* **51**, 1173–1179.

Parton, W. J., Cole, C. V., Stewart, J. W. B., Schimel, D. S., and Ojima, D. (1989a). Simulating the long-term dynamics of C, N and P in soils. *In* "Ecology of Arable Land—Perspectives and Challenges" (M. Clarholm and L. Bergstrom, eds.). Nijhoff, Dordrecht, The Netherlands..

Parton, W. J., Sanford, R. L., Sanchez, P. A., and Stewart, J. W. B. (1989b). Modeling soil organic matter dynamics in tropical soils. *In* "Dynamics of Soil Organic Matter in Tropical Ecosystems" (D. C. Coleman, J. M. Oades, and G. Uehara, eds.), pp. 153-171. NifTAL Project, Univ. of Hawaii, Honolulu.

Pastor, J., and Post, W. M. (1988). Responses of northern forests to CO_2-induced climate change. *Nature (London)* **334**, 55–58.

Pate, J. S., Layzell, D. B., and Atkins, C. A. (1979). Economy of carbon and nitrogen in a nodulated and non-nodulated (NO_3 grown) legume plant. *Plant Physiol.* **64**, 1083–1088.

Paul, E. A., and Clark, F. E. (1989). "Soil Microbiology and Biochemistry." Academic Press, San Diego.

Pawluk, S. (1987). Faunal micromorphological features in moder humus of some western Canadian soils. *Geoderma* **40**, 3–16.

Payne, W. J. (1970). Energy yields and growth of heterotrophs. *Annu. Rev. Microbiol.* **24**, 17–52.

Perdue, J. C. (1987). "Population dynamics of mites (Acari) in conventional and conservation tillage agroecosystems." Ph. D. Dissertation, University of Georgia, Athens.

Perdue, J. C., and Crossley, D. A., Jr. (1989). Seasonal abundance of soil mites (Acari) in experimetnal agroecosystems: Effects of drought in no-tillage and conventional tillage. *Soil Tillage Res.* **15**, 117–124.

Persson, T., ed. (1980). Structure and function of northern coniferous forests- an ecosystem study. *Ecol. Bull.* **32**.

Pesek, J. C., ed. (1989). "Alternative Agriculture." National Research Council Press, Washington, D.C.

Petersen, H., and Luxton, M. (1982). A comparative analysis of soil fauna populations and their role in decomposition processes. *Oikos* **39**, 287–388.

Phillips, R. E., and Phillips, S. H. (1984). "No-tillage Agriculture: Principles and Practices." Van Nostrand Reinhold, New York.

Piearce, T. G., and Phillips, M. J. (1980). The fate of ciliates in the earthworm gut: an *in vitro* study. *Microb. Ecol.* **5**, 313–320.

Pimm, S. L. (1982). "Food Webs." Chapman & Hall, London.

Pimm, S. L., and Lawton, J. H. (1980). Are food webs divided into compartments? *J. Anim. Ecol.* **49**, 879–898.

Pirozynski, K. A., and Malloch, D. W. (1975). The origin of land plants: A matter of mycotrophism. *BioSystems* **6**, 153–164.

Plass, G. N. (1956). Carbon dioxide and the climate. *Am. Sci.* **44**, 302–319.

Poinar, G. O., Jr. (1983). "The Natural History of Nematodes." Prentice Hall, Englewood Cliffs, New Jersey.

Polis, G. A. (1991). Complex trophic interactions in deserts: An empirical critique of food-web theory. *Am. Nat.* **138**, 123–155.

Pomeroy, L. R. (1974). The ocean's food web, a changing paradigm. *BioScience* **24**, 499–504.

Ponge, J. F. (1991). Food resources and diets of soil animals in a small area of Scots pine litter. *Geoderma* **49**, 33–62.

Porter, K. G. (1975). Enhancement of algal growth and productivity by grazing zooplankton. *Science* **192**, 1332–1334.

Postgate, J. R. (1987). "Nitrogen Fixation." 2nd Ed. Studies in Biology, No. 92. Arnold, London.

Postma, J., and Altemüller, H.-J. (1990). Bacteria in thin soil sections stained with the fluorescence brightener Calcofluor M2R. *Soil Biol. Biochem.* **22**, 89–96.

Powlson, D. S. (1975). Effects of biocidal treatments on soil organisms. *In* "Soil Microbiology" (N. Walker, ed.). Butterworth, London.

Powlson, D. S. (1994). The soil microbial biomass: Before, beyond and back. *In* "Beyond the Biomass" (K. Ritz, J. Dighton, and K. E. Giller, eds.), pp. 3–20. Wiley, Chichester.

Pratt, H. D., and Stjanovich, C. J. (1967). Hymenoptera. Key to some common species which sting man. *In* "Pictorial Keys to Arthropods, Reptiles, Birds and Mammals of Public Health Significance" (Anon), pp. 102–118. National Communicable Disease Center, Atlanta.

Publicover, D. A., and Vogt, K. A. (1993). A comparison of methods for estimating forest fine root production with respect to sources of error. *Can. J. For. Res.* **23**, 1179–1186.

Pussard, M., Alabouvette, C., and Levrat, P. (1994). Protozoan interactions with the soil microflora and possibilities for biocontrol of plant pathogens. *In* "Soil Protozoa" (J. Darbyshire, ed.), pp. 123–146. CAB International, Wallingford, U.K..

Read, D. J. (1991). Mycorrhizas in ecosystems. *Experientia* **47**, 376–391.

Read, D. J., Francis, R., and Finlay, R. D. (1985). Mycorrhizal mycelia and nutrient cycling in plant communities. *In* "Ecological Interactions in Soil: Plants, Microbes and Animals" (A. H. Fitter, D. Atkinson, D. J. Read, and M. B. Usher, eds.), pp. 193–217. Blackwell, Oxford.

Reeves, R. M. (1967). Seasonal distribution of some forest soil oribatei. *In* "Proceedings Second International Congress of Acarology" (G. O. Evans, ed.), pp. 23–30. Akadémiai Kiadó, Budapest.

Reeves, R. M. (1973). Oribatid ecology. *In* "Proceedings of the First Soil Microcommunities Conference." (D. L. Dindal, ed.) pp. 157–175. CONF-711-76, U. S. Atomic Energy Commission, Technical Information Center, Washington, D. C.

Reid, J. B., and Goss, M. J. (1982). Suppression of decomposition of ^{14}C-labelled plant roots in the presence of living roots of maize and perennial ryegrass. *J. Soil Sci.* **33**, 387–395.

Reynolds, J. W., and Cooke, D. G. (1993). "Nomenclatura Oligochaetologica. Supplementum Tertium." New Brunswick Museum Monographic Series No. 9. St. Johns, New Brunswick.

Reynoldson, T. B. (1939). Enchytraeid worms and the bacteria bed method of sewage treatment. *Ann. Appl. Biol.* **26**, 139–164.

Rogers, J. E., and Whitman, W. B. (1991). "Microbial Production and Consumption of Greenhouse Gasses: Methane, Nitrogen Oxides, and Halomethanes." American Society for Microbiology, Washington, D. C.

Ross, D. J., and Sparling, G. P. (1993). Comparison of methods to estimate microbial C and N in litter and soil under *Pinus radiata* on a coastal sand. *Soil Biol. Biochem.* **25**, 1591–1599.

Roth, C. H., and Joschko, M. (1991). A note on the reduction of runoff from crusted soils by earthworm burrows and artificial channels. *Z. Pflanzenerhähr. Bodenk.* **154**, 101–105.

Rothwell, F. M. (1984). Aggregation of surface mine soil by interaction between VAM fungi and lignin degradation products of lespedeza. *Plant Soil* **80**, 99–104.

Rovira, A. D., Foster, R. C., and Martin, J. K. (1979). Note on terminology: Origin, Nature and Nomenclature of the organic materials in the rhizosphere. *In* "The Soil–Root Interface" (J. L. Harley and R. S. Russell, eds.), pp. 1–4. Academic Press, London.

Rusek, J. (1975). Die bodenbildende Funktion von Collembolen und Acarina. *Pedobiologia* **15**, 299–308.

Russell, E. J. (1973). "Soil Conditions and Plant Growth." Longmans, London.

Russell, E. J., and Hutchinson, H. B. (1909). The effect of partial sterilization of soil on the production of plant food. *J. Agric. Sci.* **3**, 111–144.

St. John, T. V., and Coleman, D. C. (1983). The role of mycorrhizae in plant ecology. *Can. J. Bot.* **61**, 1005–1014.

St. John, T. V., Coleman, D. C., and Reid, C. P. P. (1983a). Association of vesicular-arbuscular mycorrhizal hyphae with soil organic particles. *Ecology* **64**, 957–959.

St. John, T. V., Coleman, D. C., and Reid, C. P. P. (1983b). Growth and spatial distribution of nutrient-absorbing organs: Selective exploitation of soil heterogeneity. *Plant Soil* **71**, 487–493.

Scharpenseel, H. W., Schomaker, M., and Ayoub, A. (1990). "Soils on a Warmer Earth." Elsevier, Amsterdam.

Scheller, U. (1990). Pauropoda. *In* "Soil Biology Guide" (D. L. Dindal, ed.), pp. 861–890. Wiley (Interscience), New York.

Schenker, R. (1984). Spatial and seasonal distribution patterns of oribatid mites (Acari: Oribatei) in a forest soil ecosystem. *Pedobiologia* **27**, 133–149.

Schimel, D. S. (1993). Population and community processes in the response of Terrestrial Ecosystems to Global Change. *In* "Biotic Interactions and Global Change" (P. M. Kareiva, J. G. Kingsolver, and R. B. Huey, eds.), pp. 45–54. Sinauer, Sunderland, Massachusetts.

Schimel, D. S., Braswell, B. H., Holland, E. A., McKeown, R., Ojima, D. S., Painter, T. H., Parton, W. J., and Townsend, A. R. (1994). Climatic, edaphic, and biotic controls over storage and turnover of carbon in soils. *Global Biogeochem. Cycles* **8**, 279–293.

Schlesinger, W. H. (1991). "Biogeochemistry. An Analysis of Global Change." Academic Press, San Diego.

Schneider, S. H. (1989). The greenhouse effect: Science and Policy. *Science* **243**, 771–781.

Schuster, R. (1956). Der anteil der Oribatiden and den Zersetzungsvorgängen im boden. *Z. Morophol. Oekol. Tiere* **45**, 1–33.

Seastedt, T. R. (1984a). Microarthropods of burned and unburned tallgrass prairie. *J. Kans. Entomol. Soc.* **57**, 468–476.

Seastedt, T. R. (1984b). The role of microarthropods in decomposition and mineralization processes. *Annu. Rev. Entomol.* **29**, 25–46.

Shamoot, S., McDonald, L., and Bartholomew, W. V. (1968). Rhizo-deposition of organic debris in soil. *Soil Sci Soc. Am. Proc.* **32**, 817–820.

Shimmel, S. M., and Darley, W. M. (1985). Productivity and density of soil algae in an agricultural system. *Ecology* **66**, 1439–1447.

Siepel, H., and de Ruiter-Dijkman, E. M. (1993). Feeding guilds of oribatid mites based on their carbohydrase activities. *Soil Biol. Biochem.* **25**, 1491–1497.

Simonson, R. W. (1959). Outline of a generalized theory of soil genesis. *Soil Sci. Soc. Am. Proc.* **23**, 152–156.

Sinclair, J. L., and Ghiorse, W. C. (1989). Distribution of aerobic bacteria, protozoa, algae, and fungi in deep subsurface sediments. *Geomicrobiol. J.* **7**, 15–31.

Singh, B. N. (1946). A method of estimating the numbers of soil Protozoa especially amoebae, based on their differential feeding on bacteria. *Ann. Appl. Biol.* **33**, 112–119.

Singh, J. S., Lauenroth, W. K., Hunt, H. W., and Swift, D. M. (1984). Bias and random errors in estimators of net root production: A simulation approach. *Ecology* **65**, 1760–1764.

Sinsabaugh, R. L., Moorhead, D. L., and Linkins, A. E. (1994). The enzymic basis of plant litter decomposition: emergence of an ecological process. *Appl. Soil Ecol.* **1**, 97–111.

Skujins, J. J. (1967). Enzymes in soil. *In* "Soil Biochemistry" (A. D. McLaren and G. H. Peterson, eds.), Vol. 1, pp. 371–414. Dekker, New York.

Smith, M. L., Bruhn, J. N., and Anderson, J. B. (1992). The root-infecting fungus *Armillaria bulbosa* may be among the largest and oldest living organisms. *Nature (London)* **356**, 428–431.

Smith, T. M., Leemans, R., and Shugart, H. H. (1992). Sensitivity of terrestrial carbon storage to CO_2-induced climate change: comparison of four scenarios based on general circulation models. *Clim. Change* **21**, 367–384.

Smucker, A. J. M., Ferguson, J. C., De Bruyn, W. P., Belford, R. K., and Ritchie, J. T. (1987). Image analysis of video-recorded plant root systems. *In* "Minirhizotron Observation Tubes: Methods and Applications for Measuring Rhizosphere Dynamics" (H. M. Taylor, ed.), pp. 67–80. American Society of Agronomy Special Publication No. 50., Madison, Wisconsin.

Snider, R. J. (1967). An annotated list of the Collembola (Springtails) of Michigan. *Michigan Entomologist* **1** 178–234.

Snider, R. J. (1969). New species of *Deuterosminthurus* and *Sminthurus* from Michigan (Collembola; Sminthuridae). *Rev. Ecol. Biol. Sol.* **3** 357–376.

Snider, R. J. (1987). Class and order Collembola. *In* "Immature Insects" (F. W. Stehr, ed.), pp. 55–64. Kendall-Hunt, Dubuque, Iowa.

Snider, R. J., Snider, R. M., and Smucker, A. J. M. (1990). Collembolan populations and root dynamics in Michigan agroecosystems. *In* "Rhizosphere Dynamics" (J. E. Box, Jr., and L. C. Hammond, eds.), pp. 168–191. Westview Press, Boulder, Colorado

Snider, R. M. (1984). Diplopoda as food for Coleoptera: Laboratory experiments. *Pedobiologia* **26**, 197–204.

Söderström, B. E. (1977). Vital staining of fungi in pure cultures and in soil with fluorescein diacetate. *Soil Biol. Biochem.* **9**, 59–63.

Southwood, T. R. E. (1978). "Ecological Methods with Particular Reference to the Study of Insect Populations," 2nd Ed. Chapman & Hall, London.

Sparling, G. P. (1981). Microcalorimetry and other methods to assess biomass and activity in soil. *Soil Biol. Biochem.* **13**, 93–98.

Staaf, H., and Berg, B. (1982). Accumulation and release of plant nutrients in decomposing scots pine needle litter. Long-term decomposition in a scots pine forest. II. *Can. J. Bot.* **60**, 1148–1168.

Stark, J. M. (1994). Causes of soil nutrient heterogeneity at different scales. *In* "Exploitation of Environmental Heterogeneity by Plants" (M. M. Caldwell and R. W. Pearcy, eds.), pp. 255–284. Academic Press, San Diego.

Steen, E. (1984). Variation of root growth in a grass ley studied with a mesh bag technique. *Swed. J. Agric. Res.* **14**, 93–97.

Steen, E. (1991). Usefulness of the mesh bag method in quantitative root studies. *In* "Plant Root Growth: An Ecological Perspective" (D. Atkinson, ed.), pp. 75–86. Blackwell, Oxford.

Stewart, J. W. B., and Cole, C. V. (1983). Influence of elemental interactions and pedogenic processes on soil organic matter dynamics. *Plant Soil* **115**, 199–209.

Stewart, J. W. B., and McKercher, R. B. (1982). Phosphorus cycle. *In* "Experimental Microbial Ecology" (R. G. Burns and J. H. Slater, eds.), pp. 221–238. Blackwell, Oxford.

Stewart, J. W. B., Anderson, D. W., Elliott, E. T., and Cole, C. V. (1990). The use of models of soil pedogenic processes in understanding changing land use and climatic change. *In* "Soils on a Warmer Earth" (H. W. Scharpenseel, M. Schomaker, and A. Ayoub, eds.), pp. 121–131. Elsevier, Amsterdam.

Stinner, B. R., and Crossley, D. A., Jr. (1980). Comparison of mineral elements cycling under till and no-till practices: An Experimental approach to agroecosystems analysis. *In* "Soil Biology as Related to Land Use Practices" (D. Dindal, ed.), pp. 180–288. U. S. Environmental Protection Agency, Washington, D.C.

Stockdill, S. M. J. (1966). The effect of earthworms on pastures. *Proc. N. Z. Ecol. Soc* **13**, 68–74.

Stork, N. E., and Eggleton, P. (1992). Invertebrates as determinants and indicators of soil quality. *Am. J. Alt. Agric.* **7**, 38–47.

Störmer, K. (1908). Ueber die Wirkung des Schwefelkohlenstoffs und aehnlicher Stoffe auf den Boden. *Zentralbl. Bakteriol.* **20**, 282–286.

Stout, J. D. (1963). The terrestrial plankton. *Tuatara* **11**, 57–65.

Stout, J. D., Goh, K. M., and Rafter, T. A. (1981). Chemistry and turnover of naturally occurring resistant organic compounds in soil. *In* "Soil Biochemistry" (E. A. Paul and J. N. Ladd, eds.), Vol. 5, pp. 1–73. Dekker, New York.

Strandtmann, R. W. (1967). Terrestrial Prostigmata (Trombidiform mites). *In* "Entomology of Antarctica." (J. L. Gressitt, ed.), Antarctic Research Series Vol. 10, pp. 51–80. American Geophysical Union, Washington, D.C.

Summerhayes, V. S., and Elton, C. S. (1923). Contributions to the ecology of Spitsbergen and Bear Island. *J. Ecol.* **11**, 214–286.

Swank, W. T., and Crossley, D. A., Jr. (1988). "Forest Hydrology and Ecology at Coweeta." Springer-Verlag, New York.

Swift, M. J., Heal, O. W., and Anderson, J. M. (1979). "Decomposition in Terrestrial Ecosystems." Univ. of California Press, Berkeley.

Syers, J. K., Sharpley, A. N., and Keeney, D. R. (1979a). Cycling of nitrogen by surface-casting earthworms in a pasture ecosystem. *Soil Biol. Biochem.* **11**, 181–185.

Syers, J. K., Springett, J. A., and Sharpley, A. N. (1979b). The role of earthworms in the cycling of phosphorus in pasture ecosystems, *Proc. Aust. Conf. Grassland Invert. Ecol.*, pp. 47–49.

Tate, K. R., and Newman, R. H. (1982). Phosphorus fractions of a climosequence of soils in New Zealand tussock grassland. *Soil Biol. Biochem.* **14**, 191–196.

Taylor, H. M. (1987). "Minirhizotron Observation Tubes: Methods and Applications for Measuring Rhizosphere Dynamics". American Society of Agronomy Special Publication No. 50, Madison, Wisconsin.

Tevis, L., Jr., and Newell, I. M. (1962). Studies on the biology and seasonal cycle of the giant red velvet mite, *Dinothrombium pandorae* (Acarina: Trombidiidae). *Ecology* **43**, 797–505.

Theng, B. K. G. (1979). "Formation and Properties of Clay–Polymer Complexes." Elsevier, Amsterdam.

Tippkötter, R., Ritz, K., and Darbyshire, J. F. (1986). The preparation of soil thin sections for biological studies. *J. Soil Sci.* **37**, 681–690.

Tisdall, J. M. (1991). Fungal hyphae and structural stability of soil. *Aust. J. Soil Res.* **29**, 729–743.

Tisdall, J. M., and Oades, J. M. (1979). Stabilization of soil aggregates by the root systems of ryegrass. *Aust. J. Soil Res* **17**, 429–441.

Tisdall, J. M., and Oades, J. M. (1982). Organic matter and waterstable aggregates in soils. *J. Soil Sci.* **33**, 141–163.

Titlyanova, A. A. (1987). Ecosystem succession and biological turnover. *Vegetatio* **50**, 43–51.

Todd, R. L., Crossley, D. A., Jr., and Stormer, J. A., Jr. (1974). "Chemical Composition of Microarthropods by Electron Microprobe Analysis: A Preliminary Report." Proc. 32nd Ann. Proc. Electron Micros. Soc. America, St. Louis, Missouri.

Tomlin, A. D. (1977). Pipeline construction—impact on soil micro- and mesofauna (Arthropoda and Annelida) in Ontario. *Proc. Entomol. Soc. Ontario* **108**, 13–17.

Torsvik, V., Salte, K., Sørheim, R., and Goksøyr, J. (1990a). Comparison of phenotypic diversity and DNA heterogeneity in a population of soil bacteria. *Appl. Environ. Microbiol.* **56**, 776–781.

Torsvik, V., Goksøyr, J., and Daae, F. L. (1990b). High diversity in DNA of soil bacteria. *Appl. Environ. Microbiol.* **56**, 782–787.

Torsvik, V., Goksøyr, J., Daae, R. L., Sørheim, R., Michalsen, J., and Salte, K. (1994). Use of DNA analysis to determine the diversity of microbial communities. *In* "Beyond the Biomass" (K. Ritz, J. Dighton, and K. E. Giller, eds.), pp. 39–49. Wiley, Chichester.

Touchot, F., Kilbertus, G., and Vannier, G. (1983). Role d'un collembole (*Folsomia candida*) au cours de la degradation des litieres de charme et de chene, en presence au en absence d'argile. *In* "New Trends in Soil Biology" (P. Lebrun, H. M. Andre, A. Demedts, C. Grégoire-Wibo, and G. Wauthy, eds.), pp. 269–280. Dieu-Brichart, Ottignies-Louvain-la-Neuve, Belgium.

Trofymow, J. A., and Coleman, D. C. (1982). The role of bacterivorous and fungivorous nematodes in cellulose and chitin decomposition in the context of a root/rhizosphere/soil conceptual model. *In* "Nematodes in Soil Systems" (D. W. Freckman, ed.), pp. 117–137. Univ. of Texas Press, Austin.

Tunlid, A., and White, D. C. (1992). Biochemical analysis of biomass, community structure, nutritional status, and metabolic activity of microbial communities in soil. *In* "Soil Biochemistry" (G. Stotzky and J.-M. Bollag, eds.), Vol. 7, pp. 229–262. Dekker, New York.

Upchurch, D. R., and Taylor, H. M. (1990). Tools for studying rhizosphere dynamics. *In* "Rhizosphere Dynamics" (J. E. Box, Jr., and L. C. Hammond, eds.), pp. 83–115. Westview Press, Boulder, Colorado.

van Breemen, N. (1992). Soil: Biotic constructions in a Gaian sense? *In* "Responses of Forest Ecosystems to Environmental Changes" (A. Teller, P. Mathy, and J. N. R. Jeffers, eds.), pp. 189–207. Elsevier Applied Science, London.

Vance, E. D., Brookes, P. C., and Jenkinson, D. S. (1987). An extraction method for measuring microbial biomass C. *Soil Biol. Biochem.* **19**, 703–707.

Vannier, G. (1973). Originalité des conditions de vie dans le sol due à la présence de l'eau: Importance thermodynamique et biologique de la porosphère. *Ann. Soc. R. Zool. Belg.* **103**, 157–167.

Vannier, G. (1981). Signification de la persistance de la pédofaune après la point de fletrissement permanent dans les sols. *Rev. Ecol. Biol. Sol* **8**, 343–365.

Vannier, G. (1987). The porosphere as an ecological medium emphasized in professor Ghilarov's work on soil animal adaptations. *Biol. Fertil. Soils* **3**, 39–44.

van Noordwijk, M., de Ruiter, P. C., Zwart, K. B., Bloem, J., Moore, J. C., van Faassen, H. G., and Burgers, S. L. G. E. (1993). Synlocation of biological activity, roots, cracks and recent organic inputs in a sugar beet field. *Geoderma* **56**, 265–276.

van Vliet, P. C. J., Beare, M. H., and Coleman, D. C. (1995). A comparison of the population dynamics and functional roles of enchytraeidae (Oligochaeta) in hardwood forest and agricultural ecosystems. *Plant Soil* **170**, 199–207.

Vargas, R., and Hattori, T. (1986). Protozoan predation of bacterial cells in soil aggregates. *FEMS Microbiol. Ecol.* **38**, 233–242.

Veeresh, G. K., and Rajagopal, D. (1983). "Applied Soil Biology and Ecology." 2nd Ed. Oxford & IBH Publ., New Delhi.

Verhoef, H. A., and De Goede, R. G. M. (1985). Effects of collembolan grazing on nitrogen dynamics in a coniferous forest. *In* "Ecological Interactions in Soil: Plants, Microbes and Animals" (A. H. Fitter, D. Atkinson, D. J. Read, and M. B. Usher, eds.), pp. 367–376. Blackwell, Oxford.

Vernadsky, V. I. (1944). Problems of biogeochemistry. II. The fundamental matter-energy difference between the living and the inert natural bodies of the biosphere. *Trans. Conn. Acad. Arts Sci.* **35**, 483–512.

Vitousek, P. M. (1994). Beyond global warming: Ecology and global change. *Ecology* **75**, 1861–1876.

Vogt, K. A., Grier, C. C., Meier, C. E., and Edmonds, R. L. (1982). Mycorrhizal role in net primary production and nutrient cycling in *Abies amabilis* ecosystems in western Washington. *Ecology* **63**, 370–380.

Vogt, K. A., Grier, C. C., Gower, S. T., Sprugel, D. G., and Vogt, D. J. (1986). Overestimation of net root production: A real or imaginary problem? *Ecology* **67**, 577–579.

Volobuev, V. R. (1964). "Ecology of Soils." Israel Program for Science Translations, Davey & Co., New York.

Voroney, R. P., and Paul, E. A. (1984). Determination of K_c and K_n *in situ* for calibration of the chloroform fumigation-incubation method. *Soil Biol. Biochem.* **16**, 9–14.

Voroney, R. P., Winter, J. P., and Gregorich, E. G. (1991). Microbe, plant, soil interactions. *In* "Carbon Isotopes Techniques" (D. C. Coleman and B. Fry, eds.), pp. 77–99. Academic Press, San Diego.

Vos, W., and Stortelder, A. H. F. (1988). Vanishing Tuscan Landscapes. Landscape ecology of a Submediterranean -Montane area (Solano Basi, Tuscany, Italy). Ph.D. Thesis. University of Amsterdam, The Netherlands.

Vossbrinck, C. R., Coleman, D. C., and Woolley, T. A. (1979). Abiotic and biotic factors in litter decomposition in a semiarid grassland. *Ecology* **60**, 265–271.

Wallace, D. F. (1994). Cat-scan assessment of earthworm (*Lumbricus terrestris* and *Lumbricus rubellus*) burrows as macropores. M.S. Thesis, University of Georgia, Athens.

Wallwork, J. A. (1970). "Ecology of Soil Animals." McGraw-Hill, London.

Wallwork, J. A. (1976). "The Distribution and Diversity of Soil Fauna." Academic Press, London.

Wallwork, J. A. (1982). "Desert Soil Fauna." Praeger, New York.

Wallwork, J. A. (1983). Oribatids in forest ecosystems. *Annu. Rev. Entomol.* **28**, 109–130.

Walsh, M. I., and Bolger, T. (1990). Effects of diet on the growth and reproduction of some Collembola in laboratory cultures. *Pedobiologia* **34**, 161–171.

Walter, D. E. (1988). Predation and mycophagy by endeostigmatid mites (Acariformes: Prostigmata). *Exp. Appl. Acarol.* **4**, 159–166.

Walter, D. E., and Ikonen, E. K. (1989). Species, guilds and functional groups: Taxonomy and behavior in nematophagous arthropods. *J. Nematol.* **21**, 315–327.

Walter, D. E., Hunt, H. W., and Elliott, E. T. (1987). The influence of prey type on the development and reproduction of some predatory soil mites. *Pedobiologia* **30**, 419–424.

Walter, D. E., Kaplan, D. T., and Permar, T. A. (1991). Missing links: A review of methods used to estimate trophic links in soil food webs. *Agric. Ecosyst. Environ.* **34**, 399–405.

Wang, G. M., Coleman, D. C., Freckman, D. W., Dyer, M. I., McNaughton, S. J., Acra, M. A., and Goeschl, J. D. (1989). Carbon partitioning patterns of mycorrhizal versus non-mycorrhizal plants. Real-time dynamic measurements using $^{11}CO_2$. *New Phytol.* **112**, 489–493.

Wardle, D. A., and Yeates, G. W. (1993). The dual importance of competition and predation as regulatory forces in terrestrial ecosystems: Evidence from decomposer food-webs. *Oecologia* **93**, 303–306.

Warnock, A. J., Fitter, A. H., and Usher, M. B. (1982). The influence of a springtail *Folsomia candida* on the mycorrhizal association of leek *Allium porrum* and arbuscular mycorrhizal endophyte *Glomus fasciculatus*. *New Phytol.* **90**, 285–292.

Waters, A. G., and Oades, J. M. (1991). Organic matter in water stable aggregates. *In* "Advances in Soil Organic Matter Research. The Impact on Agriculture and the Environment" (W. S. Wilson, ed.). Royal Society of Chemistry, Cambridge.

Webb, D. P. (1977). Regulation of deciduous forest litter decomposition by soil arthropod feces. *In* "The role of Arthropods in Forest Ecosystems" (W. J. Mattson, ed.), pp. 57–69. Springer, New York.

"Webster's New Universal Unabridged Dictionary. Deluxe Second Edition." (1983). New World Dictionaries/Simon and Schuster, Cleveland.

Whitford, W. G., Freckman, D. W., Parker, L. W., Schaefer, D., Santos, P. F., and Steinberger, Y. (1983). The contributions of soil fauna to nutrient cycles in desert systems. *In* "New Trends in Soil Biology" (P. Lebrun, H. M. André, A. de Medts, C. Grégoire-Wibo, and G. Wauthy, eds.), pp. 49–59. Dieu-Brichart, Ottignies-Louvain-la-Neuve, Belgium.

Whitney, M. (1925). "Soil and Civilization." Van Nostrand, New York.

Wieder, R. K., and Lang, G. E. (1982). A critique of the analytical methods used in examining decomposition data obtained from litter bags. *Ecology* **63**, 1636–1642.

Williams, S. C. (1987). Scorpion bionomics. *Annu. Rev. Entomol.* **32**, 275–295.

Wilson, A. T. (1978). Pioneer agriculture explosion and CO_2 levels in the atmosphere. *Nature (London)* **273**, 40–41.

Wilson, D. S. (1980). "The Natural Selection of Populations and Communities." Benjamin/Cummings, Menlo Park, California.

Wilson, K. J., Sessitsch, A., and Akkermans, A. (1994). Molecular markers as tools to study the ecology of microorganisms. *In* "Beyond the Biomass" (K. Ritz, J. Dighton, and K. E. Giller, eds.), pp. 149–156. Wiley, Chichester.

Wise, D. H. (1993). "Spiders in Ecological Webs." Cambridge Univ. Press, Cambridge.

Witkamp, M., and van der Drift, J. (1961). Breakdown of forest litter in relation to environmental factors. *Plant Soil* **15**, 295–311.

Wolters, V. (1991). Soil invertebrates—effects on nutrient turnover and soil structure—A review. *Z. Pflanzenernähr. Bodenkd.* **154**, 389–402.

Wood, T. G., Johnson, R. A., and Anderson, J. M. (1983). Modification of soils in Nigerian Savanna by soil-feeding *Cubitermes* (Isoptera, Termitidae). *Soil Biol. Biochem.* **15**, 575–579.

Woomer, P. L., and Swift, M. J. (1994). "Report of the Tropical Soil Biology and Fertility Programme." TSBF, Nairobi, Kenya.

Wright, D. H. (1988). Inverted microscope methods for counting soil mesofauna. *Pedobiologia* **31**, 409–411.

Wright, D. H., Huhta, V., and Coleman, D. C. (1989). Characteristics of defaunated soil. II. Effects of reinoculation and the role of the mineral soil. *Pedobiologia* **33**, 427–435.

Yeates, G. W. (1981). Soil nematode populations depressed in the presence of earthworms. *Pedobiologia* **22**, 191–195.

Yeates, G. W. (1988). Earthworm and enchytraeid populations in a 13-year-old agroforestry system. *N. Z. J. For. Sci.* **18**, 304–310.

Yeates, G. W., and Coleman, D. C. (1982). Role of nematodes in decomposition. *In* "Nematodes in Soil Ecosystems" (D. W. Freckman, ed.), pp. 55–80. Univ. Texas Press, Austin.

Yeates, G. W., Bongers, T., de Goede, R. G. M., Freckman, D. W., and Georgieva, S. S. (1993). Feeding habits in soil nematode families and genera-An outline for soil ecologists. *J. Nematol.* **25**, 101–313.

Zachariae, G. (1963). Was leisten Collembolen für den Waldhumus? *In* "Soil Organisms" (J. Van der Drift and J. Doeksen, eds.), pp. 109–114. . North Holland Publ. Amsterdam.

Zachariae, G. (1964). Welche Bedeutung haben Enchyträus in Waldboden? *In* "Soil Micromorphology" (A. Jongerius, ed.), pp. 57–68. Elsevier, Amsterdam.

Zachariae, G. (1965). Spuren tierischer Tätigkeit im Boden des Buchenwaldes. *Forstwissenschaftliche Forschungen* **20**, 1–68.

Zhang, B. G., Rowland, C., Lattaud, C., and Lavelle, P. (1993). Activity and origin of digestive enzymes in gut of the tropical earthworm *Pontoscolex corethrurus. Eur. J. Soil Biol.* **29**, 7–11.

Zimmerman, P. R., Greenberg, J. P., Wandiga, S. O., and Crutzen, P. J. (1982). Termites: a potentially large source of atmospheric methane, carbon dioxide and molecular hydrogen. *Science* **218**, 563–565.

Zinke, P., Stangenberger, A., Post, W., Emanuel, W., and Olson, J. (1984). Worldwide organic carbon and nitrogen data. Report ORNL/TM-8857. Oak Ridge National Laboratory, Oak Ridge, Tennesee

Zwart, K. B., and Darbyshire, J. F. (1991). Growth and nitrogenous excretion of a common soil flagellate, *Spumella* sp. *J. Soil Sci.* **43**, 145–157.

Zwart, K. B., Kuikman, P. J., and van Veen, J. A. (1994). Rhizosphere protozoa: Their significance in nutrient dynamics. *In* "Soil Protozoa" (J. F. Darbyshire, ed.), pp. 93–121. CAB International, Wallingford, U.K.

Index